Holt McDougal Mathematics

Course 1
Chapter 3 Resource Book

Copyright © Holt McDougal, a division of Houghton Mifflin Harcourt Publishing Company. All rights reserved.

Warning: No part of this work may be reproduced or transmitted in any form or by any means, electronic or mechanical, including photocopying and recording, or by any information storage or retrieval system without the prior written permission of Holt McDougal unless such copying is expressly permitted by federal copyright law.

Teachers using HOLT MCDOUGAL MATHEMATICS may photocopy complete pages in sufficient quantities for classroom use only and not for resale.

HOLT MCDOUGAL is a trademark of Houghton Mifflin Harcourt Publishing Company.

Printed in the United States of America

If you have received these materials as examination copies free of charge, Holt McDougal retains title to the materials and they may not be resold. Resale of examination copies is strictly prohibited.

Possession of this publication in print format does not entitle users to convert this publication, or any portion of it, into electronic format.

ISBN 13: 978-0-55-400727-4
ISBN 10: 0-55-400727-4

2 3 4 5   170   12 11 10

Contents

Original content Copyright © by Holt McDougal. Additions and changes to the original content are the responsibility of the instructor.

Holt McDougal Mathematics

Description of Contents

Family Involvement Pages

The Chapter Resource Book includes a set of family involvement pages for each section. The family involvement pages consist of the following items:

- **Family Letter**, a two-page letter describing the math that the student will study in each section. A list of vocabulary words from the lessons is included, and some worked-out examples are provided.
- **At-Home Practice**, a one-page worksheet with problems drawn from the content described in the Family Letter. Answers are provided on the page so that the student and his or her family can check this work at home.
- **Family Fun**, a one-page activity sheet that the student and his or her family can work on together.

Practice A, B, and C

There are three practice worksheets for every lesson. All of these reinforce the content of the lesson. Practice B is shown in the Teacher's Edition and is appropriate for the on-level student. It is also available as a workbook (Homework & Practice Workbook).

Practice A is easier than Practice B but still practices the content of the lesson. Practice C is more challenging than Practice B.

Review for Mastery

The Review for Mastery worksheet (one per lesson) provides an alternate way to teach or review the main concepts of the lesson. This worksheet is one or two pages long and is shown in the Teacher's Edition.

Challenge

The Challenge worksheet (one per lesson) enhances critical thinking skills and extends the lesson. This worksheet is shown in the Teacher's Edition.

Problem Solving

The Problem Solving worksheet (one per lesson) provides practice in problem solving and opportunities for real-world applications and for interdisciplinary connections. There are both multiple choice and short response problems. This worksheet is shown in the Teacher's Edition.

Reading Strategies

The Reading Strategies worksheet (one per lesson) provides tools to help the student master math vocabulary or symbols.

Puzzles, Twisters & Teasers

The Puzzles, Twisters & Teasers worksheet (one per lesson) provides fun practice while reinforcing the content of the lesson.

Original content Copyright © by Holt McDougal. Additions and changes to the original content are the responsibility of the instructor.

Holt McDougal Mathematics

Family Letter

3A Understanding Decimals

Dear Family,

Now that the student has an understanding of whole number applications, he or she can begin to apply that knowledge to decimals. Examples of how the student will model decimals, read and write decimals, and compare and order decimals are shown below.

Model each decimal.

0.17

1.48

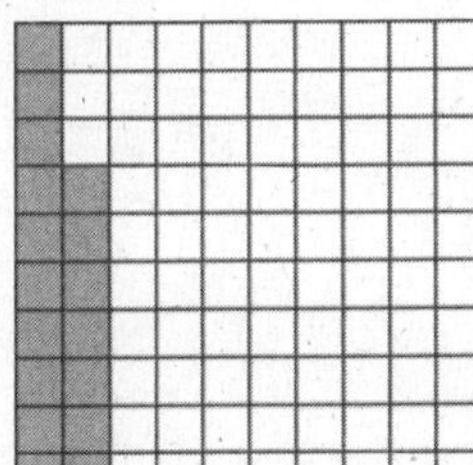
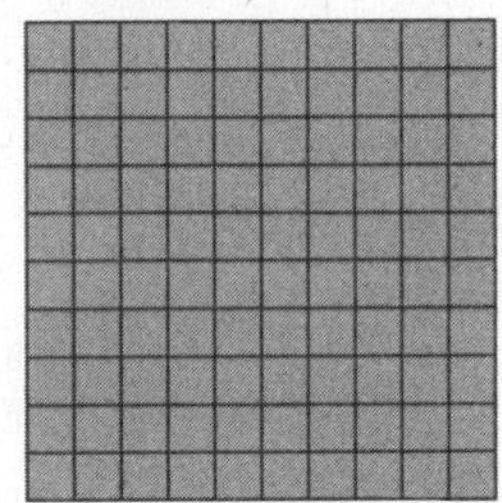
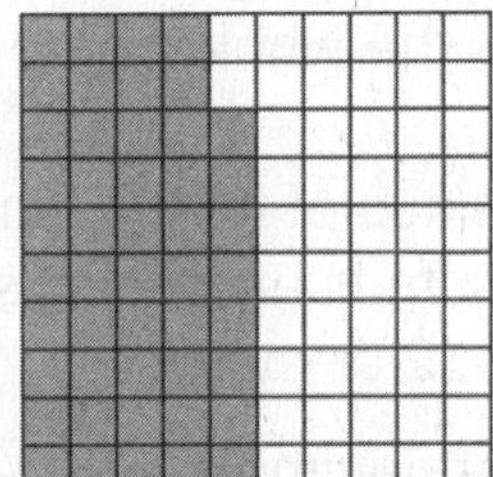

Write each decimal in standard form, expanded form, and words.

Standard Form	Expanded Form	Words
3.06	3 + 0.06	Three and six hundredths
51.128	51 + 0.128	Fifty-one and one hundred twenty-eight thousandths

Ask the student what the word *"and"* stands for when reading and writing decimals. He or she should tell you that the word *"and"* stands for the decimal point.

The student can use place value or a number line to compare and order decimals. These are the same methods the student used to compare and order whole numbers.

Compare 21.73 and 21.77.

21.73
21.77

Step 1 Line up the decimal points.

21.73
21.77

Step 2 Start at the left and compare digits.

2 1.7 3
2 1.7 7

Step 3 Look for the first place where the digits are different.

Since 7 > 3, then 21.77 > 21.73.

Vocabulary

These are the math words we are learning:

clustering rounding all of the numbers to the same value

front-end estimation estimating with only the whole-number part of the decimal

Original content Copyright © by Holt McDougal. Additions and changes to the original content are the responsibility of the instructor.

Holt McDougal Mathematics

Family Letter

3A *Understanding Decimals* continued

The student will be using estimation to find sums, differences, products, and quotients. Students may round to an indicated place value or use compatible numbers to estimate an answer to a problem. Two new estimating techniques, clustering and front-end estimation, will be introduced to the student. The following is an example of front-end estimation.

Estimate the sum using front-end estimation.
14.75 + 32.22 + 6.19

14 + 32 + 6 = 52   Add only the whole numbers.

Because the whole number values of the decimals are less than the actual numbers, the sum is an *underestimate*. Therefore the exact answer is 52 or more.

The addition and subtraction of decimals is very similar to that of whole numbers. The student will learn how to accurately add and subtract decimals by knowing where to correctly place the decimal point and when to add place-holding zeros.

Find the difference. 12.3 – 7.56

12.**30**	Align the decimal points.
– 7.56	Use **zero** as a placeholder.
4.74	Subtract. Place the decimal point.

Continue to review these decimal concepts with the student.

Sincerely,

Original content Copyright © by Holt McDougal. Additions and changes to the original content are the responsibility of the instructor.

Holt McDougal Mathematics

CHAPTER 3

At-Home Practice

3A *Understanding Decimals*

Model each decimal.

1. 0.3 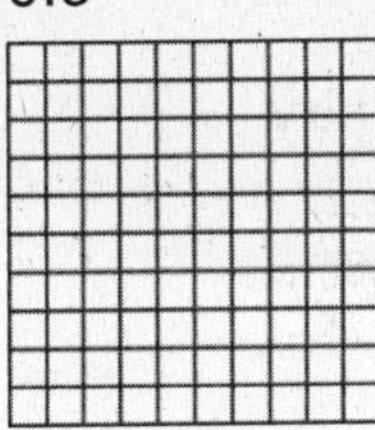

2. 0.76 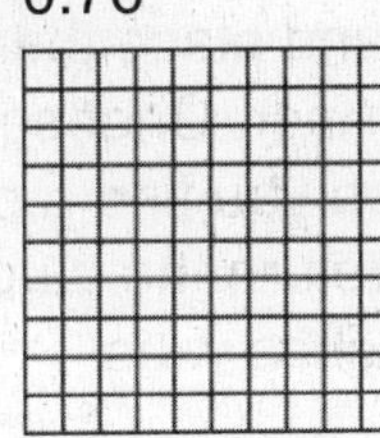

3. 1.51 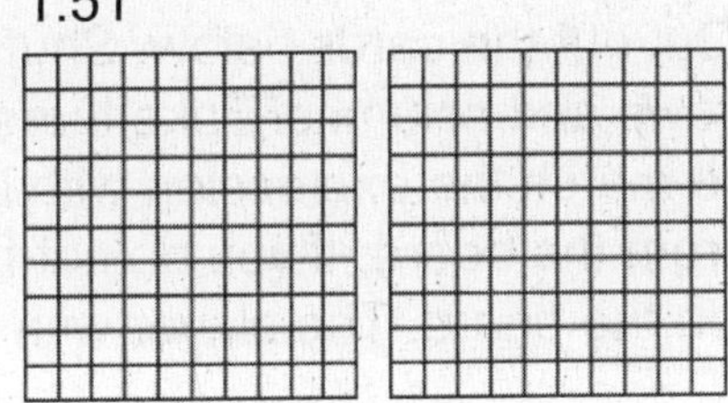

Write each decimal in standard form, expanded form, and words.

4. 3.103

5. 41 + 0.63

6. one and eight-tenths

_______________________ _______________________ _______________________

_______________________ _______________________ _______________________

Order the decimals from least to greatest.

7. 25.12, 25.07, 25.5

8. 7.33, 7.35, 7.3

_______________________ _______________________

Estimate by rounding to the indicated place value.

9. 3.0567 + 7.123; hundredths

10. 95.63 – 74.09; tenths

_______________________ _______________________

Find each sum or difference.

11. 42.18
    + 0.05

12.   5.03
    – 0.15

13. 18
    – 1.93

14. 39.12
    + 1.3

15. 8.3
    – 1.2

Answers: 1. 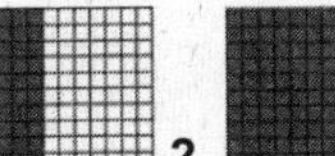 **2.**  **3.**  **4.** 3 + 0.103; three and one hundred three thousandths
5. 41.63; forty-one and sixty-three hundredths **6.** 1.8; 1 + 0.8 **7.** 25.07, 25.12, 25.5 **8.** 7.3, 7.33, 7.35 **9.** 10.18
10. 21.5 **11.** 42.23 **12.** 4.88 **13.** 16.07 **14.** 40.42 **15.** 7.1

Original content Copyright © by Holt McDougal. Additions and changes to the original content are the responsibility of the instructor.

Holt McDougal Mathematics

Name _______________________________ Date _________________ Class ____________

Family Fun
Decimal Concentration

Directions

Cut out the cards below. Shuffle the cards and place them face down in
rows and columns. Take turns with your partner. Choose two cards and try
to match the expression with its value. When you find a pair, remove it
from the board. If you do not find a match, return the cards to their spots
on the board. The player with the most matches wins.

6.02	six and 2 hundredths	10.2 − 4.35	5.85	numbers listed from greatest to least
9.602 9.6 9.062 9.006	2.089	2 + 0.089	78.9 − 51.06	about 28
12.62	twelve and sixty-two hundredths	1.06 0.98 + 10.99	about 13	204.38
204 + 0.38	44.9 + 35.12	about 90	1.0001 + 2.1	3.1001

Original content Copyright © by Holt McDougal. Additions and changes to the original content are the responsibility of the instructor.

Holt McDougal Mathematics

LESSON 3-1 Practice A

Representing, Comparing, and Ordering Decimals

Write the value of the underlined digit in each number.

1. 1.6 ________________

2. 7.62 ________________

3. 3.69 ________________

4. 20.4 ________________

5. 5.136 ________________

6. 5.08 ________________

Write each decimal in standard form, expanded form, and words.

7. 1.8 ____

8. 3 + 0.6 + 0.02 ____

9. one and fifty-two hundredths ________________________________

Circle the letter of the correct answer.

10. Which of the following sets is written in order from greatest to least?

   A 1.7, 1.07, 17

   B 5.2, 2.5, 0.52

   C 1.07, 17, 1.7

   D 2.5, 0.52, 5.2

11. Which of the following sets is written in order from least to greatest?

   F 0.85, 8.5, 5.8

   G 4.3, 3.4, 0.43

   H 5.8, 0.85, 8.5

   J 0.43, 3.4, 4.3

12. Reno, Nevada, gets an average of only five-tenths inch of rain in June, and only three-tenths inch of rain in July. Which month in Reno has less rain?

____

13. Honolulu, Hawaii, gets an average of three and eight tenths inches of rain in December, and three and six tenths inches of rain in January. Which month in Honolulu has more rain?

____

Original content Copyright © by Holt McDougal. Additions and changes to the original content are the responsibility of the instructor.

Holt McDougal Mathematics

LESSON 3-1

Practice B

Representing, Comparing, and Ordering Decimals

Write each decimal in standard form, expanded form, and words.

1. 2.07 ____

2. 5 + 0.007 ____

3. four and six tenths ___________________________________

4. sixteen and five tenths ________________________________

5. 9 + 0.6 + 0.08 _______________________________________

6. 1.037 ____

7. 2 + 0.1 + 0.003 ______________________________________

8. eighteen hundredths ___________________________________

9. 6.11 ____

Order the decimals from least to greatest.

10. 3.578, 3.758, 3.875

____

11. 0.0943, 0.9403, 0.9043

____

12. 12.97, 12.957, 12.75

____

13. 1.09, 1.901, 1.9, 1.19

____

14. Your seventh and eighth ribs are two of the longest bones in your body. The average seventh rib is nine and forty-five hundredths inches long, and the average eighth rib is 9.06 inches long. Which bone is longer?

____

15. The average female human heart weighs nine and three tenths ounces, while the average male heart weighs eleven and one tenth ounces. Which human heart weighs less, the male or the female?

____

16. The state has $42.3 million for a new theater. The theater that an architect designed would cost $42.25 million. Can the theater be built for the amount the state can pay?

____

17. Lyn traveled 79.47 miles on Saturday, 54.28 miles on Sunday, 65.5 miles on Monday, and 98.43 miles on Tuesday. Which day did she travel the greatest number of miles?

____

Original content Copyright © by Holt McDougal. Additions and changes to the original content are the responsibility of the instructor.

Holt McDougal Mathematics

LESSON 3-1 Practice C
Representing, Comparing, and Ordering Decimals

Compare. Write <, >, or =.

1. 10.569 _____ $10 + 0.05 + 0.006 + 0.0009$

2. seventy-five hundredths _____ 7.50

3. twenty thousandths _____ twelve hundredths

4. 98.30675 _____ $90 + 8 + 0.03 + 0.007$

Order from least to greatest.

5. $12.8962, 12.9682, 12.8692$ _______________________

6. $8.098, 7.098, 8.079, 7.089$ _______________________

7. $65.21, 6.521, 6.0521, 65.12$ _______________________

8. $0.304, 0.30, 0.403, 0.43, 0.34$ _______________________

Order from greatest to least.

9. $9.653, 90.563, 90.6053$ _______________________

10. $11.717, 11.771, 11.117, 11.171$ _______________________

11. $8.0359, 8.3509, 8.359, 8.5$ _______________________

12. $2.35, 2.05, 2.03, 2.30, 2.53, 2.5$ _______________________

13. From 1984 to 1996 American Carl Lewis won every men's gold medal for the long jump at the Summer Olympic Games. In 1984, he jumped 8.54 meters. In 1988, he jumped 8.72 meters. In 1992, he jumped 8.67 meters, and in 1996, he jumped 8.5 meters. Write Lewis' long jumps in order from least to greatest distance. In which year did he jump the farthest?

____

14. Some of the previous world records for the highest pole vaults are 6.01 meters, 6.03 meters, 6.0 meters, and 6.14 meters. Write the records in order from greatest to least height. What is the world record for the highest pole vault?

____

Original content Copyright © by Holt McDougal. Additions and changes to the original content are the responsibility of the instructor.

Holt McDougal Mathematics

LESSON 3-1	**Review for Mastery**

Review for Mastery
Representing, Comparing, and Ordering Decimals

You can use place value to write decimals
in standard form, expanded form, and word form.

To write 2.14 in expanded form, write the decimal
as an addition expression using the place value
of each digit.

2.14 can be written as $2 + 0.1 + 0.04$.

When you write a decimal in word form, the number
before the decimal point tells you how many wholes
there are. The decimal point stands for the word "and."

Notice that the place value names to the right of the decimal begin
with tenths, hundredths, and then thousandths. The "ths" ending
indicates a decimal.

2.14 can also be written as *two and fourteen hundredths.*

Ones	Tenths	Hundredths	Thousandths	Ten Thousandths
2	1	4		

1. How would you read a number with 4 decimal places?

____

Write each decimal in standard form, expanded form, and word form.

2.

Ones	Tenths	Hundredths	Thousandths	Ten Thousandths
5	6	9	8	

____

____

____

3.

Ones	Tenths	Hundredths	Thousandths	Ten Thousandths
0	0	9	4	

____

____

____

4. $7 + 0.8$

____

____

5. twelve-hundredths

Original content Copyright © by Holt McDougal. Additions and changes to the original content are the responsibility of the instructor.

Holt McDougal Mathematics

Review for Mastery
LESSON 3-1
Representing, Comparing, and Ordering Decimals (cont.)

You can use place value to compare decimals.
Use < or > to compare the decimals.

Ones	Tenths	Hundredths	Thousandths	Ten Thousandths
3	7	6	8	
3	7	5	4	

0.06 > 0.05, so 3.768 > 3.754.

Compare. Write >, <, or =.

6.

Ones	Tenths	Hundredths	Thousandths	Ten Thousandths
1	0	3		
1	3			

1.03 ____ 1.3

7.

Ones	Tenths	Hundredths	Thousandths	Ten Thousandths
4	6	7		
4	6	7	0	

4.67 ____ 4.670

8.

Ones	Tenths	Hundredths	Thousandths	Ten Thousandths
0	3	6	4	5
0	3	4	6	5

0.3645 ____ 0.3465

9. 8.53 ____ 8.053

10. 2.253 ____ 2.1345

11. 0.87 ____ 0.08703

You can use place value to order decimals.

To order 9.76, 8.59, and 9.24, from least to greatest,
first compare the numbers in pairs.

Ones	Tenths	Hundredths	Thousandths	Ten Thousandths
9	7	6		
8	5	9		
9	2	4		

9.76 > 8.59, 8.59 < 9.24, 9.76 > 9.24.

So the numbers from least to greatest are 8.59, 9.24, 9.76.

Order the decimals from least to greatest.

12. 0.54, 0.43, 0.52

13. 3.43, 3.34, 3.4

14. 8.9, 9.8, 9.5

____

____

____

15. 0.83, 0.8, 0.083

16. 1.1, 0.01, 1.01

17. 6.5, 6.0, 0.6

____

____

____

Original content Copyright © by Holt McDougal. Additions and changes to the original content are the responsibility of the instructor.

Holt McDougal Mathematics

LESSON 3-1

Challenge

Place Your Values

Complete the tables below to show different numbers that can be written with the same digits. Do not use the same digit more than once for each place value.

1. Use the digits 1, 3, 5, 7, and 9 to write four 5-digit numbers of increasing value.

Hundreds	Tens	Ones	Tenths	Hundredths	Thousandths	Ten-Thousandths

2. Use the digits 0, 2, 4, 6, 7, and 8 to write four 6-digit numbers of decreasing value.

Hundreds	Tens	Ones	Tenths	Hundredths	Thousandths	Ten-Thousandths

3. Use the digits 0, 1, 2, 3, 4, 5, and 6 to write four 7-digit numbers of increasing value.

Hundreds	Tens	Ones	Tenths	Hundredths	Thousandths	Ten-Thousandths

Original content Copyright © by Holt McDougal. Additions and changes to the original content are the responsibility of the instructor.

   Holt McDougal Mathematics

LESSON 3-1 — Problem Solving
Representing, Comparing, and Ordering Decimals

Use the table to answer the questions.

Largest Marine Mammals

Mammal	Length (ft)	Weight (T)
Blue whale	110.0	127.95
Fin whale	82.0	44.29
Gray whale	46.0	32.18
Humpback whale	49.2	26.08
Right whale	57.4	39.37
Sperm whale	59.0	35.43

1. What is the heaviest marine mammal on Earth?

   ____

2. Which mammal in the table has the shortest length?

   ____

3. Which mammal in the table is longer than a humpback whale, but shorter than a sperm whale?

   ____

Circle the letter of the correct answer.

4. Which mammal measures forty-nine and two tenths feet long?

   A   blue whale

   B   gray whale

   C   sperm whale

   D   humpback whale

5. Which mammal weighs thirty-five and forty-three hundredths tons?

   F   right whale

   G   sperm whale

   H   gray whale

   J   fin whale

6. Which of the following lists shows mammals in order from the least weight to the greatest weight?

   A   sperm whale, right whale, fin whale, gray whale

   B   fin whale, sperm whale, gray whale, blue whale

   C   fin whale, right whale, sperm whale, gray whale

   D   gray whale, sperm whale, right whale, fin whale

7. Which of the following lists shows mammals in order from the greatest length to the least length?

   F   sperm whale, right whale, humpback whale, gray whale

   G   gray whale, humpback whale, right whale, sperm whale

   H   right whale, sperm whale, gray whale, humpback whale

   J   humpback whale, gray whale, sperm whale, right whale

Original content Copyright © by Holt McDougal. Additions and changes to the original content are the responsibility of the instructor.

Holt McDougal Mathematics

LESSON 3-1

Reading Strategies
Connect Symbols and Words

You can read and write decimals in three ways. A place value chart can help you read decimals.

When you read or say a decimal, say "and" when you come to the decimal point.

Ones	Tenths	Hundredths
2 . 5		
0 . 1	7	
8 . 0	6	

Read:

2 **and** 5 tenths

17 hundredths

8 **and** 6 hundredths

Use this chart to help you write decimals in standard form and in expanded form.

Words and Symbols	Standard Form	Expanded Form
2 and 5 tenths	2.5	2 + 0.5
17 hundredths	0.17	0.1 + 0.07
8 and 6 hundredths	8.06	8 + 0.06

Write each decimal in words and symbols, standard form, or expanded form.

1. Write 2.17 with words and symbols. _______________________________

2. Write 2.17 in expanded form. _______________________________

3. Write 3 and 6 hundredths in standard form. _______________________________

4. Write 3 and 6 hundredths in expanded form. _______________________________

5. Write 1.5 with words and symbols. _______________________________

6. Write 1.5 in expanded form. _______________________________

Original content Copyright © by Holt McDougal. Additions and changes to the original content are the responsibility of the instructor.

Holt McDougal Mathematics

LESSON 3-1 Puzzles, Twisters & Teasers

Criss-Cross Riddle

For each decimal written in standard form in the left column, there is
a matching number in the right column. Draw a line connecting the
numbered circles on the left to the circles beside their matches on
the right. To solve the riddle, find the letters crossed by the lines from
each number and write the letters with that corresponding number in
the blanks below.

Left		Letters	Right
7.251	(1)	T	Nine and sixty-five hundredths
		P	
2.5	(2)	Y	Twelve and one hundredth
		E	
65.65	(3)	K	7 + 0.2 + 0.05 + 0.001
		O	
9.65	(4)	G	3 + 0.2 + 0.01 + 0.002
		L	
3.212	(5)	I	four and four tenths
		N	
12.01	(6)	F	2 + 0.5
4.4	(7)		sixty-five and sixty-five hundredths

Riddle:

What is another name for a parrot that has flown away?

A __ __ __ __ __ __ __ !
    1   2   3   4   5   6   7

Original content Copyright © by Holt McDougal. Additions and changes to the original content are the responsibility of the instructor.

  Holt McDougal Mathematics

LESSON 3-2	**Practice A**
	Estimating Decimals

Round each decimal to the underlined place value.

1. 1.<u>7</u>8

2. 0.5<u>6</u>9

3. 12.<u>6</u>2

4. 3.<u>2</u>15

5. 24.6<u>0</u>8

6. <u>3</u>7.84

Estimate by rounding to the indicated place value.

7. 3.67 + 1.23; tenths

8. 0.726 + 0.119; hundredths

9. 12.86 − 5.73; tenths

10. 8.643 − 2.795; nearest whole number

Estimate each product or quotient.

11. 17.6 ÷ 6.2

12. 1.9 • 7.045

13. 23.8 ÷ 4.3

14. 9.02 • 4.65

15. 36.1 ÷ 3.9

16. 2.8 • 5.35

17. Latoya measured the growth of a plant for her science project. When she started the project, the plant was 2.8 inches tall. At the end of the project, the plant was 5.2 inches tall. About how many inches did the plant grow during Latoya's project?

18. Tyler bought 16.2 yards of cloth to make costumes for the school play. He needs 3.8 yards of the cloth to make each costume. About how many costumes can Tyler make with the cloth he bought?

Original content Copyright © by Holt McDougal. Additions and changes to the original content are the responsibility of the instructor.

Holt McDougal Mathematics

Practice B

LESSON 3-2

Estimating Decimals

Estimate by rounding to the indicated place value.

1. 7.462 + 1.809; tenths

2. 15.3614 − 2.0573; hundredths

3. 56.4059 − 4.837; ones

4. 0.60871 + 1.2103; hundredths

Estimate each product or quotient.

5. 42.1 ÷ 5.97

6. 11.8 • 6.125

7. 63.78 ÷ 8.204

8. 7.539 • 3.0642

9. 80.794 ÷ 8.61

10. 19.801 • 2.78

Estimate a range for each sum.

11. 6.8 + 4.3 + 5.6

12. 12.63 + 9.86 + 20.30

13. Two sixth-grade classes are collecting money to buy a present for one of their teachers. One class collected $24.68 and the other class collected $30.25. About how much money did they collect in all? The gift they want to buy costs $69.75. About how much more money do they need?

14. On the highway, Anita drove an average speed of 60.2 miles per hour. At that speed, about how far can she travel in three and a half hours? At that same speed, about how many hours will it take Anita to drive 400 miles?

Original content Copyright © by Holt McDougal. Additions and changes to the original content are the responsibility of the instructor.

Holt McDougal Mathematics

LESSON 3-2	**Practice C**

Practice C
Estimating Decimals

Estimate by rounding to the indicated place value.

1. 102.68 + 57.209; ones

2. 83.56912 − 11.7415; hundredths

3. 215.673 − 9.5511; tenths

4. 99.97452 + 5.6409; tenths

Estimate each product or quotient.

5. 24.674 • 5.578

6. 107.95 ÷ 9.193

7. 72.603 ÷ 10.164

8. 6.694 • 38.69

9. 140.9 ÷ 11.756

10. 48.6035 • 7.63

Estimate a range for each sum.

11. 8.4 + 3.1 + 9.54 + 2.86

12. 17.38 + 5.93 + 18.60 + 29.34

13. For the first half of the year, the average monthly precipitation levels in Atlanta, Georgia, are 4.8 inches, 4.8 inches, 5.8 inches, 4.3 inches, 4.3 inches, and 3.6 inches. Estimate the average precipitation level in Atlanta for January through June. Then find the actual average precipitation level for those months. How does your estimate compare to the actual data?

14. Tammy earns $14.68 per hour working as a chef. On average she works 40.3 hours a week. About how much does Tammy earn each week? About how much does she earn each year?

Original content Copyright © by Holt McDougal. Additions and changes to the original content are the responsibility of the instructor.

Holt McDougal Mathematics

Review for Mastery

LESSON 3-2

Estimating Decimals

You can use rounding to estimate. Round to the indicated place value.
Then add or subtract.

A. 3.478 + 7.136; tenths

   3.478   7 ≥ 5, so round up   3.5
   7.136   3 < 5, so round down +7.1
                                   10.6

   3.478 + 7.136 is about 10.6.

B. 12.848 – 6.124; hundredths

   12.848   8 ≥ 5, so round up   12.85
    6.124   4 < 5, so round down –6.12
                                   6.73

   12.848 – 6.124 is about 6.73.

Estimate by rounding to the indicated place value.

1. 1.04 + 9.37; tenths

   1.04 rounds to _______

   9.37 rounds to _______

   estimate _______

2. 2.17 + 3.56; tenths

   2.17 rounds to _______

   3.56 rounds to _______

   estimate _______

3. 6.753 – 4.245; hundredths

   6.753 rounds to _______

   4.255 rounds to _______

   estimate _______

You can use compatible numbers to estimate. Pick numbers that are
close to the actual numbers that are easy to multiply or divide. Then
multiply or divide.

A. 4.6 • 3.2
   5 and 3 are compatible numbers.
   5 • 3 = 15, so 4.6 • 3.2 is about 15.

B. 48.3 ÷ 13.2
   48 and 12 are compatible numbers.
   48 ÷ 12 = 4, so 48.3 ÷ 13.2 is
   about 4.

Use compatible numbers to estimate each product or quotient.

4. 9.4 • 5.6

5. 7.25 • 10.84

6. 84.8 ÷ 3.9

7. 21.9 ÷ 3.1

_______   _______   _______   _______

8. 8.3 • 7.6

9. 55.7 ÷ 6.9

10. 5.57 ÷ 2.7

11. 6.729 • 9.8

_______   _______   _______   _______

Original content Copyright © by Holt McDougal. Additions and changes to the original content are the responsibility of the instructor.

 Holt McDougal Mathematics

Challenge

Out to Lunch

**Use the restaurant bills below to estimate the total cost of each
meal. Then estimate the amount each person should pay to
split each check evenly.**

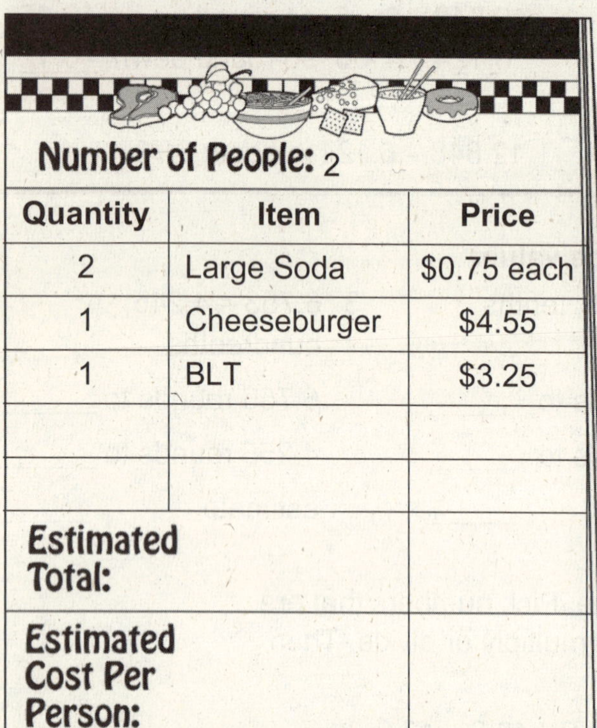

Number of People: 2

Quantity	Item	Price
2	Large Soda	$0.75 each
1	Cheeseburger	$4.55
1	BLT	$3.25
Estimated Total:		
Estimated Cost Per Person:		

Number of People: 4

Quantity	Item	Price
2	Slice of Pie	$1.45 each
1	Brownie	$1.25
3	Hot Tea	$0.60 each
Estimated Total:		
Estimated Cost Per Person:		

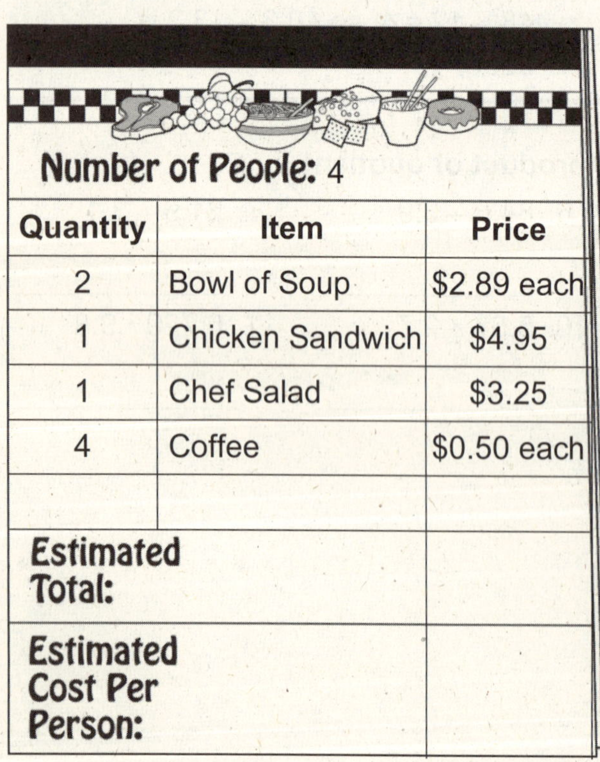

Number of People: 4

Quantity	Item	Price
2	Bowl of Soup	$2.89 each
1	Chicken Sandwich	$4.95
1	Chef Salad	$3.25
4	Coffee	$0.50 each
Estimated Total:		
Estimated Cost Per Person:		

Number of People: 5

Quantity	Item	Price
1	Pizza	$14.95
5	House Salad	$2.85 each
4	Large Soda	$0.75 each
1	Small Soda	$0.55
Estimated Total:		
Estimated Cost Per Person:		

Original content Copyright © by Holt McDougal. Additions and changes to the original content are the responsibility of the instructor.

Holt McDougal Mathematics

<table><tr><td>LESSON
3-2</td><td></td></tr></table>

Problem Solving
Estimating Decimals

Write the correct answer.

1. Men in Iceland have the highest average life expectancy in the world—76.8 years. The average life expectancy for a man in the United States is 73.1 years. About how much higher is a man's average life expectancy in Iceland? Round your answer to the nearest whole year.

2. The average life expectancy for a woman in the United States is 79.1 years. Women in Japan have the highest average life expectancy—3.4 years higher than the United States. Estimate the average life expectancy of women in Japan. Round your answer to the nearest whole year.

3. There are about 1.6093 kilometers in one mile. There are 26.2 miles in a marathon race. About how many kilometers are there in a marathon race? Round your answer to the nearest tenths.

4. At top speed, a hornet can fly 13.39 miles per hour. About how many hours would it take a hornet to fly 65 miles? Round your answer to the nearest whole number.

Circle the letter of the correct answer.

5. The average male human brain weighs 49.7 ounces. The average female human brain weighs 44.6 ounces. What is the difference in their weights?

   A about 95 ounces

   B about 7 ounces

   C about 5 ounces

   D about 3 ounces

6. An official hockey puck is 2.54 centimeters thick. About how thick are two hockey pucks when one is placed on top of the other?

   F about 4 centimeters

   G about 4.2 centimeters

   H about 5 centimeters

   J about 5.2 centimeters

7. Lydia earned $9.75 per hour as a lifeguard last summer. She worked 25 hours a week. About how much did she earn in 8 weeks?

   A about $250.00

   B about $2,000.00

   C about $2,500.00

   D about $200.00

8. Brent mixed 4.5 gallons of blue paint with 1.7 gallons of white paint and 2.4 gallons of red paint to make a light purple paint. About how many gallons of purple paint did he make?

   F about 9 gallons

   G about 8 gallons

   H about 10 gallons

   J about 7 gallons

Original content Copyright © by Holt McDougal. Additions and changes to the original content are the responsibility of the instructor.

Holt McDougal Mathematics

Reading Strategies

LESSON 3-2

Use Context

You estimate to get an approximate answer. Rounding decimals to the nearest whole number is one way to estimate.

Mike's mom bought 3.28 pounds of cheddar cheese. She also bought 2.75 pounds of Swiss cheese. About how many pounds of cheese did she buy?

To round to the nearest whole number, look at the tenths place.

    3.**28** ←— 2 is less than 5; round down to 3.
    +2.**75** ←— 7 is greater than 5; round up to 3.

3.28 + 2.75 rounded to the nearest whole number is:

    ↓   ↓
    3 + 3 = 6 pounds of cheese.

Complete each problem.

1. Which decimal place value do you look at to round to the nearest whole number? _________________________

2. Round 34.67 pounds to the nearest pound. _________________________

3. Round 42.19 pounds to the nearest pound. _________________________

4. Estimate this sum: 42.19 pounds + 34.67 pounds. _________________________

5. Round $54.14 to the nearest dollar. _________________________

6. Round $21.54 to the nearest dollar. _________________________

7. Estimate the difference: $54.14 – $21.54. _________________________

Original content Copyright © by Holt McDougal. Additions and changes to the original content are the responsibility of the instructor.

Holt McDougal Mathematics

LESSON 3-2 — Puzzles, Twisters & Teasers

Secret Message

Recently you've noticed that your neighbor Slippery Larry has been extremely forgetful when he leaves his house. He keeps leaving, then returning to get all of the things he forgot. One day you decided to keep track of Larry by measuring how many feet he walked down his driveway before returning to the house. Here's what you found:

1.	Keys	6.435	**8.**	Soccer Ball	13.575
2.	Wallet	4.909	**9.**	Laptop	18.249
3.	Jacket	4.899	**10.**	Suspenders	13.59
4.	Pants	14.642	**11.**	Rope	1.59
5.	Watch	12.94	**12.**	Gloves	14.253
6.	Briefcase	4.851	**13.**	Hat	18.161
7.	Pen	1.491			

It just so happens that Slippery Larry was on the news later that day. He is actually a criminal communicating to his criminal buddies through a code system. Seems his friends watch how far he travels down his driveway and the distances translate into the following code:

1.6	A	19.0	H	9.4	O	13.0	V
3.6	B	21.2	I	8.8	P	13.8	W
8.4	C	14.8	J	9.6	Q	1.4	X
4.4	D	9.2	K	10.8	R	1.2	Y
5.0	E	2.6	L	18.2	S	5.2	Z
6.8	F	6.4	M	13.6	T		
15.6	G	14.6	N	14.4	U		

Since your measurements are more exact than Slippery Larry's and his friends', you'll have to round each of your distances to match their code. Then write the appropriate letters next to your rounded distance. Finally, write these letters above the appropriate blanks below to discover where you can catch Slippery Larry! (Numbers under the blanks refer to numbers in bold from your list).

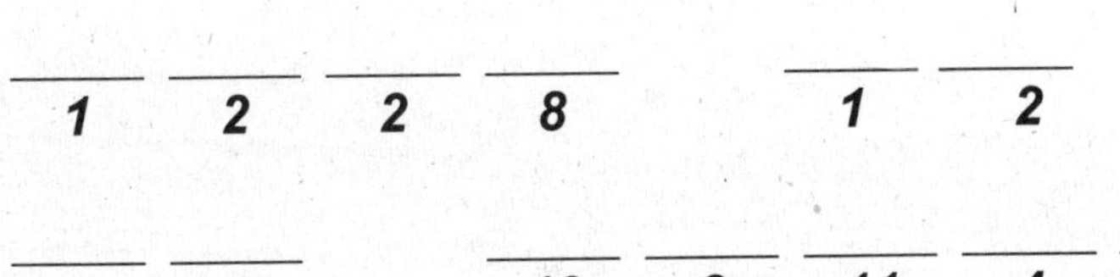

___ ___ ___ ___   ___ ___
 1   2   2   8   1   2

___ ___   ___ ___ ___ ___ ___
11   8   13   8   11   4   13

Original content Copyright © by Holt McDougal. Additions and changes to the original content are the responsibility of the instructor.

LESSON
3-3

Practice A
Adding and Subtracting Decimals

Find each sum or difference.

1. $1.5 + 2.3$

2. $6.5 + 1.4$

3. $8.9 - 5.1$

4. $12.6 - 3.4$

5. $8.16 - 7.02$

6. $7.25 + 8.75$

7. $11.4 + 8.6$

8. $16.5 - 4.3$

9. $9.55 - 1.2$

10. $25.6 + 5.1$

11. $8.9 + 3.05$

12. $10.64 - 8.5$

Circle the letter of the correct answer.

13. If $x = 2.3$, what is the value of the expression $5.4 + x$?

   A 3.1   C 7.1

   B 7.7   D 3.7

14. If $a = 4.2$, what is the value of the expression $8.7 - a$?

   F 12.9   H 4.5

   G 4.9   J 12.5

15. If $m = 1.9$, what is the value of the expression $m + 4.2$?

   A 2.3   C 6.1

   B 2.2   D 7.1

16. If $y = 5.9$, what is the value of the expression $7.2 - y$?

   F 1.3   H 13.3

   G 1.7   J 13.1

17. Marcus is 1.5 meters tall. His sister, Carol, is 0.1 meter taller than Marcus. Their father is 0.2 meter taller than Carol. How tall is Carol? How tall is their father?

18. Jennifer brought $14.75 to the baseball game. She spent $3.45 for a hot dog and soda. How much money does she have left?

Original content Copyright © by Holt McDougal. Additions and changes to the original content are the responsibility of the instructor.

Holt McDougal Mathematics

LESSON 3-3

Practice B
Adding and Subtracting Decimals

Find each sum or difference.

1. $8.9 + 2.4$

2. $12.7 - 9.6$

3. $18.35 - 4.16$

4. $7.21 + 11.6$

5. $0.975 + 3.8$

6. $20.66 - 9.1$

7. Tiffany's job requires a lot of driving. How many miles did she travel during the month of February? ________________

Miles Tiffany Traveled

Week	1	2	3	4
Miles	210.05	195.18	150.25	165.30

8. Shelly babysits after school and on the weekends. How much did she earn in all for the month of April? ________________

Shelly's Earnings for April

Week	1	2	3	4
Earnings	$120.50	$180.75	$205.25	$215.50

Evaluate $5.6 - a$ for each value of a.

9. $a = 3.7$

10. $a = 0.5$

11. $a = 2.8$

12. $a = 1.42$

13. $a = 0.16$

14. $a = 3.75$

15. Allen bought a box of envelopes for $2.79 and a pack of paper for $4.50. He paid with a $10 bill. How much change should he receive?

16. From a bolt of cloth measuring 25.60 yards, Tina cut a 6.8-yard piece and an 11.9-yard piece. How much material is left on the bolt?

Original content Copyright © by Holt McDougal. Additions and changes to the original content are the responsibility of the instructor.

Holt McDougal Mathematics

LESSON 3-3

Practice C
Adding and Subtracting Decimals

Find each sum or difference.

1. $7.9 + 1.38$

2. $100 - 65.9$

3. $204.965 + 55.8$

4. $63.057 - 18.45$

5. $0.5541 - 0.09$

6. $11.2398 + 8.9 + 2.7$

7. $0.22 + 15.607 + 9.7$

8. $20 - 3.78 - 0.64$

9. $1.9 + 0.25 + 9.4 + 0.5$

Evaluate each expression.

10. $39.702 - a$ for $a = 0.9$

11. $x + 1.064$ for $x = 28.5$

12. $50.02 - p$ for $p = 0.99$

13. $w - 7.08$ for $w = 100$

14. $37.62 + t$ for $t = 8.084$

15. $u + 2.7$ for $u = 0.5046$

Write the missing digit in each problem.

16.
$$\begin{array}{r} 7.089 \\ +\ 2.\square13 \\ \hline 9.502 \end{array}$$

17.
$$\begin{array}{r} 16.594 \\ -\ \square.175 \\ \hline 11.419 \end{array}$$

18.
$$\begin{array}{r} 6.2\square67 \\ +\ 9.75 \\ \hline 15.9867 \end{array}$$

19.
$$\begin{array}{r} 0.6\square9 \\ -\ 0.458 \\ \hline 0.221 \end{array}$$

20.
$$\begin{array}{r} 238.793 \\ +\ 19.5\square5 \\ \hline 258.305 \end{array}$$

21.
$$\begin{array}{r} 100.\square5 \\ -\ 19.99 \\ \hline 80.46 \end{array}$$

22. Italian Delight sells three sizes of pizzas at different prices. If you buy all three pizzas, it costs a total of $46.24. A medium pizza costs $15.75, and a large costs $17.50. How much does a small pizza cost?

23. Brent has three sheets of plywood that are each 6.85 feet long. He cut a 3.4-foot piece from one sheet and a 0.5-foot piece from the other. How many feet of plywood does he have left in all?

Original content Copyright © by Holt McDougal. Additions and changes to the original content are the responsibility of the instructor.

Holt McDougal Mathematics

LESSON 3-3

Review for Mastery
Adding and Subtracting Decimals

You can use a place-value chart to help you add and subtract decimals.

Add 1.4 and 0.9.

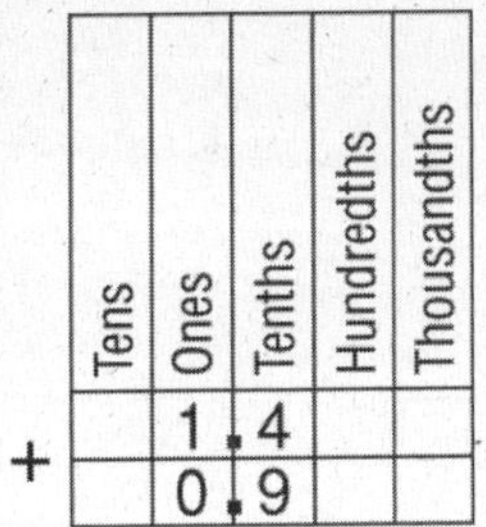

So, 1.4 + 0.9 = 2.3.

Subtract 2.4 from 3.1.

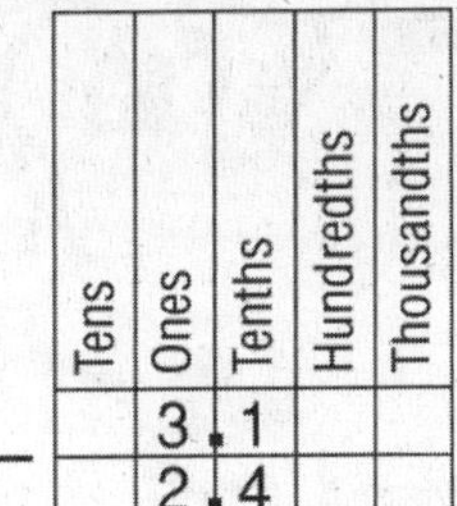

So, 3.1 − 2.4 = 0.7.

Find each sum or difference.

1.

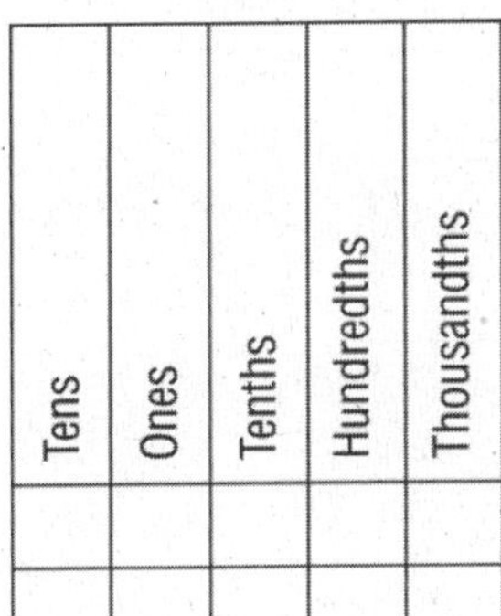

____

2.

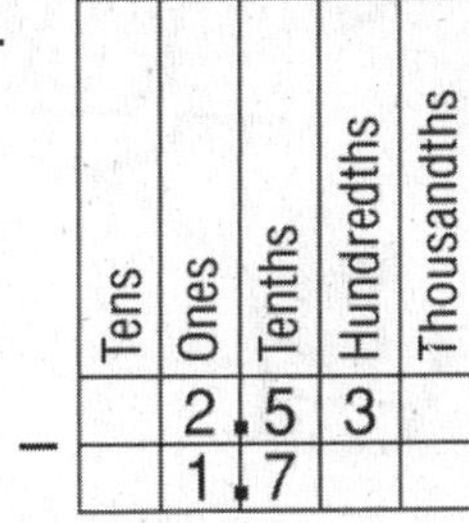

____

3. 4.3 + 1.4

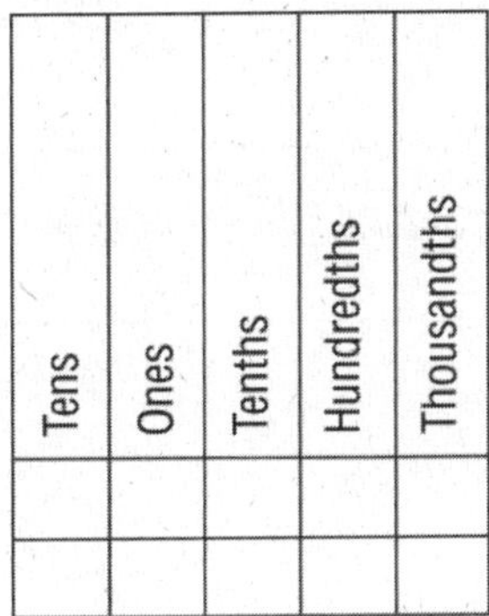

____

4. 14.4 − 3.8

____

5. 7.3 + 8.5

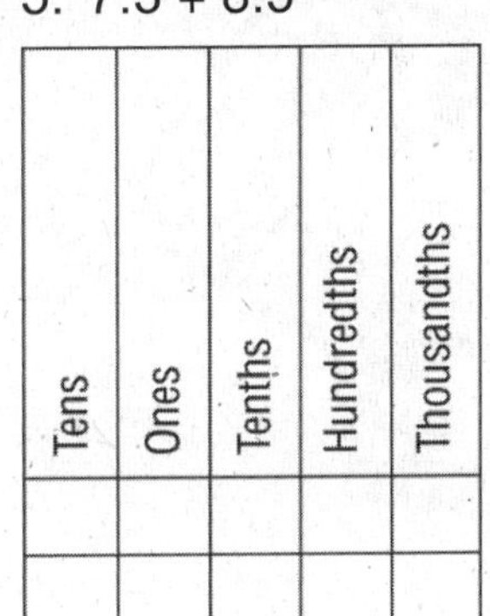

____

6. 12.34 − 6.9

____

7. 6.3 − 2.5

____

8. 20.65 + 13.24

____

9. 8.9 − 1.95

____

10. 3.42 + 5.25

____

Original content Copyright © by Holt McDougal. Additions and changes to the original content are the responsibility of the instructor.

Holt McDougal Mathematics

LESSON 3-3

Challenge

A Penny Saved Is a Penny Earned

Next to each bank, describe three different coin combinations that equal the amount of money it holds. For each combination, use at least one quarter, one dime, one nickel, and one penny.

1.

____

____

____

2.
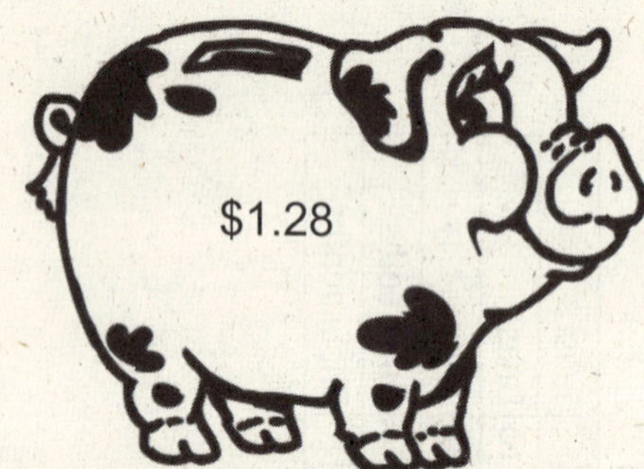

____

____

____

3.

____

____

____

4.

____

____

____

Original content Copyright © by Holt McDougal. Additions and changes to the original content are the responsibility of the instructor.

LESSON 3-3

Problem Solving

Adding and Subtracting Decimals

Use the table to answer the questions.

Busiest Ports in the United States

Port	Imports Per Year (millions of tons)	Exports Per Year (millions of tons)
South Louisiana, LA	30.6	57.42
Houston, TX	75.12	33.43
New York, NY & NJ	53.52	8.03
New Orleans, LA	26.38	21.73
Corpus Christi, TX	52.6	7.64

1. How many more tons of imports than exports does the Port of New Orleans handle each year?

2. How many tons of imports and exports are shipped through the port of Houston, Texas, each year in all?

____

____

Circle the letter of the correct answer.

3. Which port ships 0.39 more tons of exports each year than the port at Corpus Christi, Texas?

    A Houston

    B NY & NJ

    C New Orleans

    D South Louisiana

4. What is the difference between the imports and exports shipped in and out of Corpus Christi's port each year?

    F 45.04 million tons

    G 44.94 million tons

    H 44.96 million tons

    J 44.06 million tons

5. What is the total amount of imports shipped into the nation's 5 busiest ports each year?

    A 238.22 million tons

    B 366.47 million tons

    C 128.25 million tons

    D 109.97 million tons

6. What is the total amount of exports shipped out of the nation's 5 busiest ports each year?

    F 366.47 million tons

    G 128.25 million tons

    H 109.97 million tons

    J 238.22 million tons

Original content Copyright © by Holt McDougal. Additions and changes to the original content are the responsibility of the instructor.

Holt McDougal Mathematics

LESSON 3-3	**Reading Strategies**
	Use an Organizer

Writing decimals in a place-value grid helps you line up decimal points to add or subtract decimals.

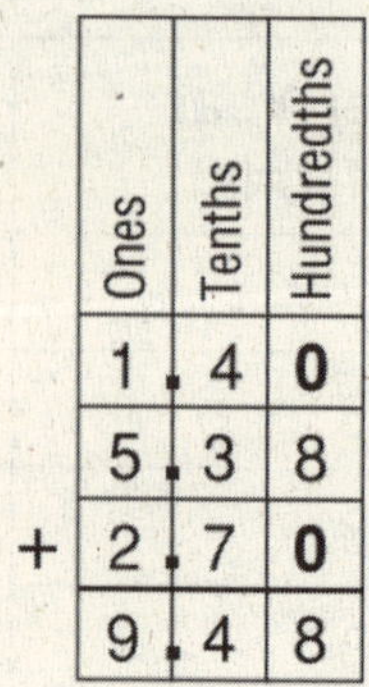

```
  1.40
  5.38
+2.70
  9.48
```

Add zeros as place holders.

Place decimal point in answer.

```
  28.05
-  6.30
  21.75
```

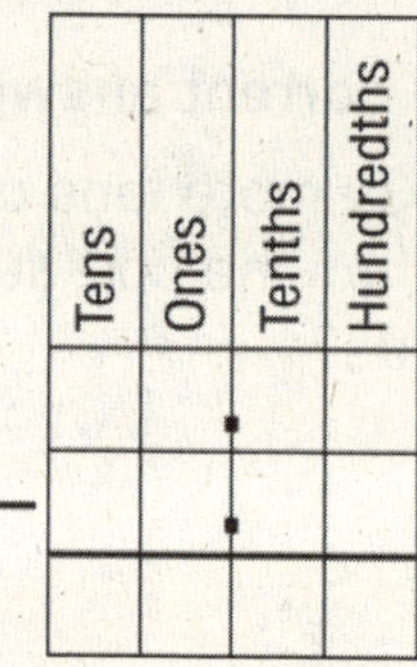

1. How does the place-value grid help you add or subtract?

____

2. Place these numbers on the place-value grid below: 3.25, 1.06, 2.9.

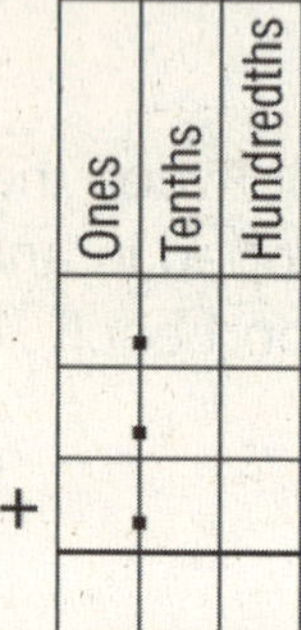

3. Place this problem on the place-value grid below: 23.82 – 7.2.

4. Add the numbers on the place-value grid. What is the sum?

____

____

5. Subtract the numbers on the place-value grid. What is the difference?

____

____

6. For which numbers did you add zero as a place holder?

____

____

Original content Copyright © by Holt McDougal. Additions and changes to the original content are the responsibility of the instructor.

Holt McDougal Mathematics

LESSON 3-3

Puzzles, Twisters & Teasers

Crazy Pete's Letter Shop

What's the quickest way to double your money?

To solve this riddle, you need to buy your letters from Crazy Pete.

Right now, you know three letters from the first word: **L, F, O.** You also know three letters from the second word: **F, A, H.**

To use those six letters in the riddle, you need to buy them from Crazy Pete.

You have $15.00 to spend on each word, but you know you need to buy two more letters to complete the riddle—one for each word.

Crazy Pete tells you a secret: "You have exactly enough money left after buying the first three letters of each word to purchase the missing letter."

Using the price table at right, add the cost of the three letters you need to buy for the first word. Subtract this total from $15.00 to find the cost of the missing letter.

Then do it again for the second word.

First Word Cost (letters **L, F, O**): _______________________

Remaining money: _______________________

Missing Letter: _______________________

Second Word Cost (letters **F, A, H**): _______________________

Remaining money: _______________________

Final Letter: _______________________

A	$1.50
B	$5.43
C	$6.08
D	$4.13
E	$1.50
F	$6.08
G	$4.79
H	$2.83
I	$1.50
J	$6.08
K	$4.79
L	$4.59
M	$5.43
N	$3.29
O	$0.20
P	$4.79
Q	$7.87
R	$3.29
S	$3.29
T	$3.29
U	$1.50
V	$4.79
W	$5.43
X	$7.87
Y	$5.43
Z	$7.87

Unscramble the letters of each word to solve the riddle!

______ ______ ______ ______ it in

______ ______ ______ ______ !

Original content Copyright © by Holt McDougal. Additions and changes to the original content are the responsibility of the instructor.

Holt McDougal Mathematics

Family Letter

3B Multiplying and Dividing Decimals

Dear Family,

The student is continuing to learn about decimals and decimal applications. One application involves using decimals to express large numbers. This method is called scientific notation. An example is shown below.

Write 6,358,000 in scientific notation.

Step 1   Move the decimal point left to form a number greater than one but less than ten.

6,358,000.0

6.358000

$$6.358000 > 1$$
$$6.358000 < 10$$

Step 2   Since the decimal point moved 6 places to the left, the number is multiplied by ten to the sixth power.

Step 3   Write the number in scientific notation.
$$6,358,000 = 6.358 \times 10^6$$

The student will also learn how to write a number in standard form when given a number written in scientific notation.

Write 7.82×10^5 in standard form.

7.82×10^5 The power of 10 is 5.

7.82000   Move the decimal point 5 places to the right.

$7.82 \times 10^5 = 782,000$.

The student is learning to multiply a decimal by a whole number and a decimal by another decimal. He or she is also learning to divide a decimal by a whole number and a decimal by a decimal. These skills will help the student evaluate products and quotients for given variables and learn to solve decimal equations.

The following page gives examples of the steps the student will use to multiply and divide decimals. Notice that the process is similar to the one used to multiply and divide whole numbers except that you have to place a decimal point in the product or quotient.

Vocabulary

These are the math words we are learning:

scientific notation
a shorthand method that can be used to write numbers

Original content Copyright © by Holt McDougal. Additions and changes to the original content are the responsibility of the instructor.

Holt McDougal Mathematics

Family Letter

3B *Multiplying and Dividing Decimals* continued

Multiply a Decimal by a Whole Number

6×0.4

$6 \times 0.4 \rightarrow$ Multiply as you would with whole numbers.

$6 \times 4 = 24$

Add the number of decimal places in each number multiplied.

$6 \quad \rightarrow \quad 0$ decimal place

$0.4 \rightarrow \underline{+ \, 1 \text{ decimal place}}$

$\qquad\qquad$ **1** decimal place

$6 \times 0.4 = 2.4 \rightarrow$ Place the decimal point **1** digit to the left.

Multiply a Decimal by a Decimal

0.6×0.4

$0.6 \times 0.4 \rightarrow$ Multiply as you would with whole numbers.

$6 \times 4 = 24$

Add the number of decimal places in each number multiplied.

$0.6 \rightarrow \quad 1$ decimal place

$0.4 \rightarrow \underline{+ \, 1 \text{ decimal place}}$

$\qquad\qquad$ **2** decimal places

$0.6 \times 0.4 = 0.24 \rightarrow$ Place the decimal point **2** digits to the left.

Divide a Decimal by a Whole Number

$15.81 \div 3$

Place the decimal point in the quotient directly above the decimal point in the dividend.

$$
\begin{array}{r}
5.27 \\
3\overline{)15.81} \\
-15 \\
\hline
8 \\
-6 \\
\hline
21 \\
-21 \\
\hline
0
\end{array}
$$
Divide as you would with whole numbers.

Divide a Decimal by a Decimal

$15.81 \div 0.03$

Make the divisor, 0.03, a whole number by multiplying it and the dividend, 15.81, by the same power of ten.

$0.03 \times 100 = 3$

$15.81 \times 100 = 1{,}581$

$$
\begin{array}{r}
527 \\
3\overline{)1581} \\
-15 \\
\hline
8 \\
-6 \\
\hline
21 \\
-21 \\
\hline
0
\end{array}
$$
Divide

It is important that the student understand the reasonableness of an answer when performing decimal operations. When solving equations with decimal coefficients, have the student explain the answers to make certain that the solutions are reasonable.

Sincerely,

Original content Copyright © by Holt McDougal. Additions and changes to the original content are the responsibility of the instructor.

Holt McDougal Mathematics

CHAPTER 3

At-Home Practice

3B Multiplying and Dividing Decimals

Write each number in scientific notation.

1. 67,003 _________

2. 974,875 _________

3. 1,000,000 _________

Write each number in standard form.

4. 3.8×10^4 _________

5. 2.8×10^2 _________

6. 7.82×10^6 _________

Find each product.

7.
$$\begin{array}{r} 0.3 \\ \times\ 0.6 \\ \hline \end{array}$$

8.
$$\begin{array}{r} 45 \\ \times\ 0.4 \\ \hline \end{array}$$

9.
$$\begin{array}{r} 2.71 \\ \times\ 8.0 \\ \hline \end{array}$$

Write and solve an equation.

10. A sheet of drywall is on sale for $2.89. How much will 25 sheets of drywall cost?

____

Evaluate 13x for each value of x.

11. $x = 0.35$

12. $x = 1.51$

13. $x = 7.8$

Evaluate $x \div 8$ for each value of x.

14. $x = 0.824$

15. $x = 40.6$

16. $x = 16.016$

Find each quotient.

17. $7.2 \div 1.2$

18. $71.4 \div 0.03$

19. $68.76 \div 1.8$

Solve.

20. The soccer team raised $260.75 for new uniforms. Each uniform costs $26. How many new uniforms can the soccer team purchase?

____

21. There are 22 students in Mrs. Field's classroom. If juice boxes come in packs of 4, how many packs does Mrs. Field need for her class?

____

Answers: 1. 6.7003×10^4 **2.** 9.74875×10^5 **3.** 1×10^6 **4.** 38,000 **5.** 280 **6.** 7,820,000 **7.** 0.18 **8.** 18 **9.** 21.68 **10.** $72.25 **11.** 4.55 **12.** 19.63 **13.** 101.4 **14.** 0.103 **15.** 5.075 **16.** 2.002 **17.** 6 **18.** 2,380 **19.** 38.2 **20.** 10 uniforms **21.** 6 packs

Original content Copyright © by Holt McDougal. Additions and changes to the original content are the responsibility of the instructor.

Holt McDougal Mathematics

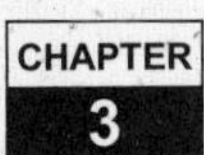

CHAPTER 3

Family Fun
Capture the Flag

Directions

Solve each problem. Then find the sum of the digits of your answer. Shade the square if the sum is an odd number. The picture made by the shaded squares will tell you under which block the flag is. If you evaluate each number correctly, you will capture your enemy's flag.

22×2.6	0.67×10^3	$8.7 \div 0.3$	12×10^2
20×10^5	18.2×5.1	0.5×0.4	3.21×7
$10.55 \div 5$	$9.2 \div 0.5$	6.81×3	$27.6 \div 100$
1.5×0.5	$19 \div 0.2$	33×100	$98 \div 0.7$
9.8×0.4	$1.1 \div 0.5$	$18.3 \div 6$	0.6×0.3

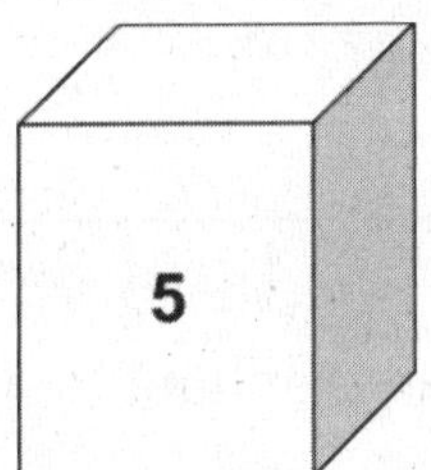

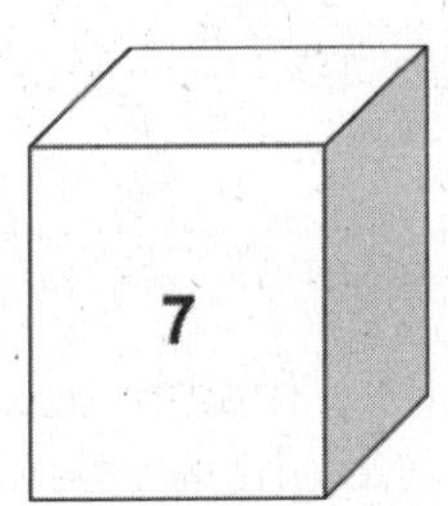

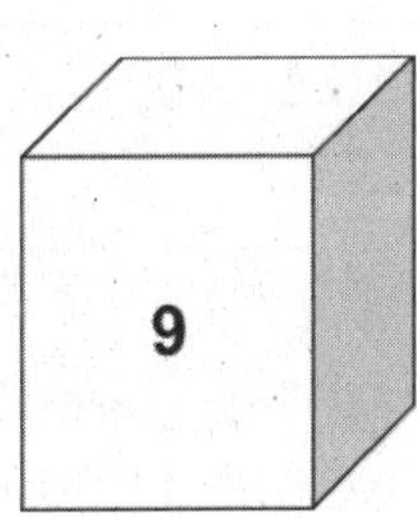

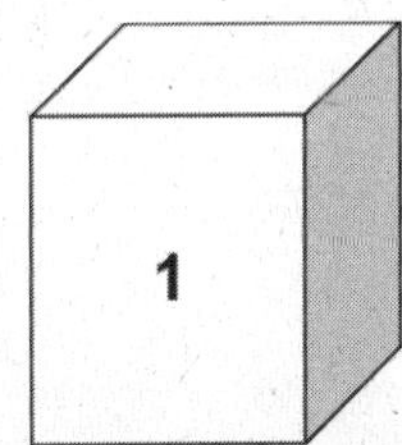

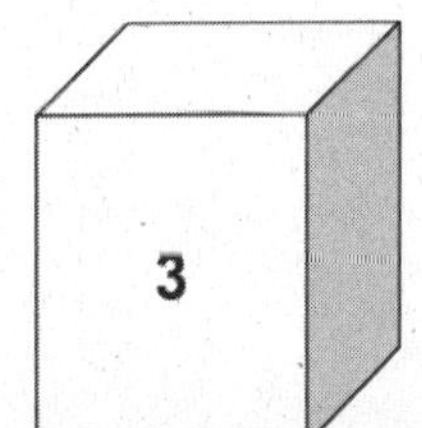

Answer: The flag is under the 9 block.

Original content Copyright © by Holt McDougal. Additions and changes to the original content are the responsibility of the instructor.

Holt McDougal Mathematics

LESSON 3-4

Practice A
Scientific Notation

Find each product.

1. 267 • 100

2. 38.1 • 100

3. 1.92 • 100

Circle the letter of the correct answer.

4. Which of the following shows 85,000 written in scientific notation?

   A $8.5 \cdot 10^3$

   B $8.5 \cdot 10^4$

   C $8.5 \cdot 10^5$

   D $8.5 \cdot 10^6$

5. Which of the following shows $3.67 \cdot 10^5$ written in standard form?

   F 3,670

   G 36,700

   H 367,000

   J 3,670,000

Fill in the blanks to make each equation true.

6. $1,200 = 1.2 \cdot 10^{\underline{}}$

7. $25,000 = 2.5 \cdot 10^{\underline{}}$

8. $580 = 5.8 \cdot 10^{\underline{}}$

9. $470,000 = \underline{} \cdot 10^5$

10. $6,580 = \underline{} \cdot 10^3$

11. $8,900,000 = \underline{} \cdot 10^6$

Write each number in standard form.

12. $3.4 \cdot 10^2$

13. $7.9 \cdot 10^4$

14. $1.75 \cdot 10^3$

15. $1.24 \cdot 10^5$

16. $9.6 \cdot 10^5$

17. $1.28 \cdot 10^6$

18. African elephants are the largest land mammals. The average African elephant weighs 11,000 pounds. Write this weight in scientific notation.

Original content Copyright © by Holt McDougal. Additions and changes to the original content are the responsibility of the instructor.

Holt McDougal Mathematics

Practice B

LESSON 3-4

Scientific Notation

Find each product.

1. $345 \cdot 100$

2. $65.2 \cdot 100$

3. $1.84 \cdot 1{,}000$

Write each number in scientific notation.

4. 16,700

5. 4,680

6. 58,340,000

Write each number in standard form.

7. $3.25 \cdot 10^4$

8. $7.08 \cdot 10^6$

9. $1.209 \cdot 10^7$

10. $6.8 \cdot 10^8$

11. $0.51 \cdot 10^5$

12. $0.006 \cdot 10^3$

Identify the answer choice that is _not_ equal to the given number.

13. 356,000

   A $300{,}000 + 56{,}000$

   B $3.56 \cdot 10^5$

   C $3.56 \cdot 10^4$

14. $1.28 \cdot 10^6$

   A $100{,}000 + 28{,}000$

   B $1{,}280{,}000$

   C $12.8 \cdot 10^5$

15. 1,659,000

   A $1{,}600{,}000 + 59{,}000$

   B $1.659 \cdot 10^6$

   C $16.59 \cdot 10^6$

16. $0.074 \cdot 10^3$

   A $70.0 + 4.0$

   B $7.4 \cdot 10^5$

   C $7.4 \cdot 10^1$

17. In 2000, the population of Pennsylvania was 12,281,054. Round this figure to the nearest hundred thousand. Then write that number in scientific notation.

18. In 2000, the population of North Carolina was about $8.05 \cdot 10^6$, and the population of South Carolina was about $4.01 \cdot 10^6$. Write the combined populations of these two states in standard form.

Original content Copyright © by Holt McDougal. Additions and changes to the original content are the responsibility of the instructor.

Holt McDougal Mathematics

LESSON 3-4

Practice C

Scientific Notation

Find each product.

1. $1.67 \times 1{,}000$

2. 93.6×100

3. $3.55 \times 10{,}000$

Write each number in scientific notation.

4. 6,389,000

5. 105,200,000

6. 152 million

Write each number in standard form.

7. $1.5089 \cdot 10^4$

8. $2.516 \cdot 10^8$

9. $1.7711 \cdot 10^7$

10. $3.9604 \cdot 10^6$

11. $0.284 \cdot 10^4$

12. $0.0869 \cdot 10^2$

Write each measurement using scientific notation.

13. 250 km = _______________ m

14. 0.065 kg = _______________ g

15. 89 L = _______________ mL

16. 1,540 km = _______________ cm

17. 0.73 m = _______________ mm

18. 10,240 kg = _______________ g

19. In a recent count, 147,171,000 people in the United States owned cars. In the same year, $4.268 \cdot 10^7$ people in Japan owned cars. In which country did more people own cars? How many more?

20. On average, about $1.1 \cdot 10^9$ passengers use the New York City subway system each year. About 1,170,000,000 passengers use the Paris subway each year. How many passengers use those two subways each year?

Original content Copyright © by Holt McDougal. Additions and changes to the original content are the responsibility of the instructor.

Holt McDougal Mathematics

LESSON
3-4

Review for Mastery

Scientific Notation

Scientific notation expresses a large number as the product of a number between one and ten and a power of ten.

To write 3,400 in scientific notation, move the decimal point to the left until the number falls between 1 and 10.

3,400 $1 < 3 < 10$, so move the decimal point 3 places to the left.

$3,400 = 3.4 \cdot 10^3$ The number of times you move the decimal point left is the power of ten.

Write each number in scientific notation.

1. 175,000

2. 298

3. 5,764

4. 83

5. 40,300

6. 2,000,000

7. 51,010

8. 190,025

You can express numbers written in scientific notation in standard form.

The power of ten tells you how many places to move the decimal point to the right.

$3.2 \cdot 10^4 = 32,000$ To write $3.2 \cdot 10^4$ in standard form, move the decimal point 4 places to the right.

Write each number in standard form.

9. $5.62 \cdot 10^3$

10. $7.238 \cdot 10^2$

11. $9.9 \cdot 10^5$

12. $6.53 \cdot 10^1$

13. $5.36 \cdot 10^4$

14. $2.4 \cdot 10^2$

15. $4.35 \cdot 10^3$

16. $8 \cdot 10^5$

17. $1 \cdot 10^4$

18. $2.03 \cdot 10^3$

19. $1.12 \cdot 10^2$

20. $3.002 \cdot 10^6$

Original content Copyright © by Holt McDougal. Additions and changes to the original content are the responsibility of the instructor.

Holt McDougal Mathematics

Challenge

LESSON 3-4

The Solar System

Write the average distance from the Sun in standard form. Then use the distances to label our solar system shown below.

		Average Distance From the Sun (mi)	
		Scientific Notation	**Standard Form**
1.	Earth	$9.29 \cdot 10^7$	
2.	Jupiter	$4.836 \cdot 10^8$	
3.	Mars	$1.416 \cdot 10^8$	
4.	Mercury	$3.6 \cdot 10^7$	
5.	Neptune	$2.794 \cdot 10^9$	
6.	Pluto	$3.675 \cdot 10^9$	
7.	Saturn	$8.87 \cdot 10^8$	
8.	Uranus	$1.784 \cdot 10^9$	
9.	Venus	$6.72 \cdot 10^7$	

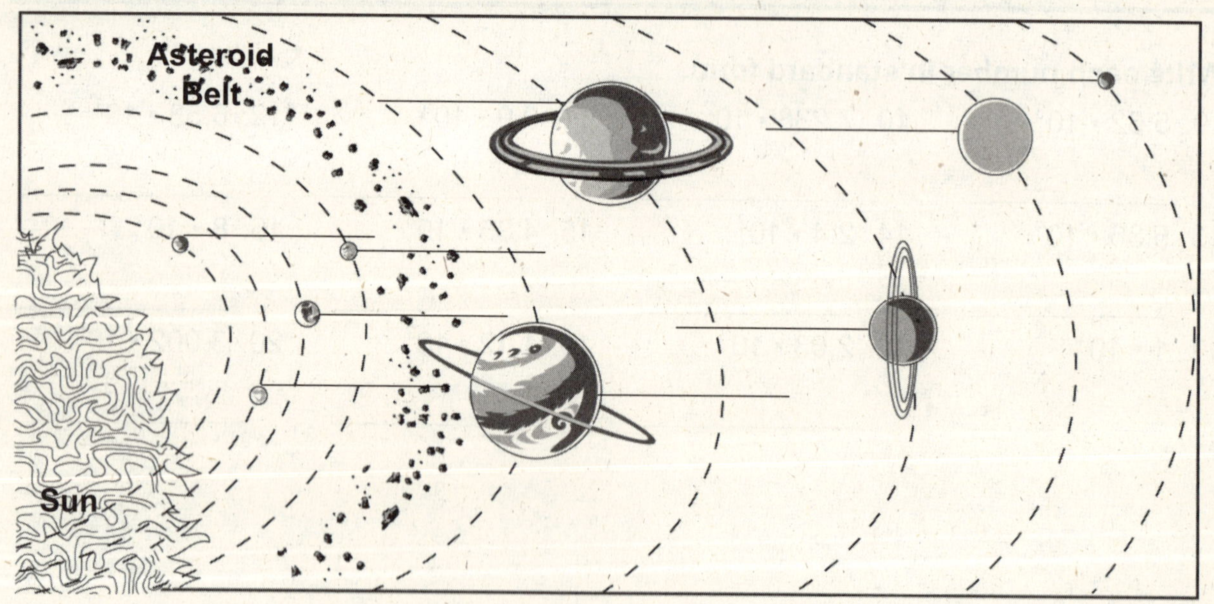

Original content Copyright © by Holt McDougal. Additions and changes to the original content are the responsibility of the instructor.

Holt McDougal Mathematics

LESSON 3-4
Problem Solving
Scientific Notation

Write the correct answer.

1. The closest comet to approach Earth was called Lexell. On July 1, 1770, Lexell was observed about 874,200 miles from Earth's surface. Write this distance in scientific notation.

2. Scientists estimate that it would take $1.4 \cdot 10^{10}$ years for light from the edge of our universe to reach Earth. How many years is that written in standard form?

3. In the United States, about 229,000,000 people speak English. About 18,000,000 people speak English in Canada. Write in scientific notation the total number of English speaking people in the United States and Canada.

4. South Africa is the top gold-producing country in the world. Each year it produces $4.688 \cdot 10^{8}$ tons of gold! Written in standard form, how many tons of gold does South African produce each year?

Circle the letter of the correct answer.

5. About $3.012 \cdot 10^{6}$ people visit Yellowstone National Park each year. What is that figure written in standard form?

   A 30,120,000 people

   B 3,012,000 people

   C 301,200 people

   D 30,120 people

6. In 2000, farmers in Iowa grew 1,740,000 bushels of corn. What is this amount written in scientific notation?

   F $1.7 \cdot 10^{5}$

   G $1.74 \cdot 10^{5}$

   H $1.74 \cdot 10^{6}$

   J $1.74 \cdot 10^{7}$

7. The temperature at the core of the Sun reaches 27,720,000°F. What is this temperature written in scientific notation?

   A $2.7 \cdot 10^{7}$

   B $2.72 \cdot 10^{7}$

   C $2.772 \cdot 10^{6}$

   D $2.772 \cdot 10^{7}$

8. Your body is constantly producing red blood cells—about $1.73 \cdot 10^{11}$ cells a day. How many blood cells is that written in standard form?

   F 173,000,000 cells

   G 17,300,000,000 cells

   H 173,000,000,000 cells

   J 1,730,000,000,000 cells

Original content Copyright © by Holt McDougal. Additions and changes to the original content are the responsibility of the instructor.

Holt McDougal Mathematics

LESSON 3-4

Reading Strategies
Use a Graphic Organizer

This chart helps you see the ways large numbers can be written.

Scientific Notation
- A number between 1 and 10 multiplied by a power of 10
  2.5×10^6
  7×10^5

Ways to Write Large Numbers

Standard Form
- All place values are shown.
  2,500,000
  700,000

Words and Symbols
- Use numbers and words.
  2 million, 500 thousand; 2.5 million
  700 thousand

Use the graphic organizer to answer Exercises 1–2.

1. Which way to write large numbers shows every place value?

   ____

2. Which way to write large numbers uses a power of 10?

   ____

Identify how each large number is written. Write "scientific notation," "standard form," or "words and symbols."

3. 8,296,000

   ____

4. 3.6 million

   ____

5. 2.9×10^5

   ____

Original content Copyright © by Holt McDougal. Additions and changes to the original content are the responsibility of the instructor.

   Holt McDougal Mathematics

LESSON 3-4 — Puzzles, Twisters & Teasers

The Scientific Two-Step

There are two simple steps to do the Scientific Two-Step and
answer the riddle:

1. For each number in scientific notation in the left column, find the same number in standard notation in the key on the right.

2. Fill in the letter associated with the standard notation answer to solve the riddle.

Key:

1. 2.458×10^1
2. 1.276×10^8
3. 2.458×10^2
4. 1.276×10^6
5. 2.458×10^4
6. 2.458×10^9
7. 1.276×10^7
8. 2.458×10^3
9. 2.458×10^5
10. 1.276×10^3
11. 2.458×10^7
12. 1.276×10^4
13. 1.276×10^2

12,760	E
127,600,000	O
24,580,000	D
245.8	M
24,580	A
2,458	F
12,760,000	Y
1,276	L
24.58	C
245,800	O
1,276,000	P
2,458,000,000	N
127.6	D

Why did the people working at the blanket factory lose their jobs?

The ___ ___ ___ ___ ___ ___ ___
   1   2   3   4   5   6   7

___ ___ ___ ___ ___ ___
8   9 10 11 12 13

Original content Copyright © by Holt McDougal. Additions and changes to the original content are the responsibility of the instructor.

     Holt McDougal Mathematics

LESSON 3-5

Practice A
Multiplying Decimals

Find each product.

1. 0.4
    $\underline{\times\ 0.2}$

2. 0.3
    $\underline{\times\ 0.4}$

3. 1.2
    $\underline{\times\ 0.5}$

4. 1.1
    $\underline{\times\ 0.9}$

5. 2.5
    $\underline{\times\ 0.5}$

6. 6.0
    $\underline{\times\ 0.7}$

7. $0.4 \cdot 0.5$

8. $1.2 \cdot 1.5$

9. $1.7 \cdot 0.3$

10. $6.7 \cdot 0.4$

11. $9.6 \cdot 0.2$

12. $0.8 \cdot 0.8$

Evaluate 2x for each value of *x*.

13. $x = 0.1$

14. $x = 0.5$

15. $x = 0.9$

16. $x = 1.2$

17. $x = 1.7$

18. $x = 2.4$

19. Each box can hold 2.5 pounds of apples. How many pounds can 3 boxes hold?

20. Each pie costs $5.60. How much will it cost to buy 2 pies?

Original content Copyright © by Holt McDougal. Additions and changes to the original content are the responsibility of the instructor.

Holt McDougal Mathematics

LESSON 3-5

Practice B

Multiplying Decimals

Find each product.

1. $\begin{array}{r} 0.7 \\ \times\ 0.3 \\ \hline \end{array}$

2. $\begin{array}{r} 0.05 \\ \times\ 0.4 \\ \hline \end{array}$

3. $\begin{array}{r} 8.0 \\ \times\ 0.02 \\ \hline \end{array}$

4. $\begin{array}{r} 3.5 \\ \times\ 0.2 \\ \hline \end{array}$

5. $\begin{array}{r} 12.1 \\ \times\ 0.01 \\ \hline \end{array}$

6. $\begin{array}{r} 9.0 \\ \times\ 0.9 \\ \hline \end{array}$

7. $0.04 \cdot 0.58$

8. $2.15 \cdot 1.5$

9. $1.73 \cdot 0.8$

10. $6.017 \cdot 2.0$

11. $3.96 \cdot 0.4$

12. $0.7 \cdot 0.009$

Evaluate 8x for each value of x.

13. $x = 0.5$

14. $x = 2.3$

15. $x = 0.74$

16. $x = 3.12$

17. $x = 0.587$

18. $x = 14.08$

19. The average mail carrier walks 4.8 kilometers in a workday. How far do most mail carriers walk in a 6-day week? There are 27 working days in July, so how far will a mail carrier walk in July?

20. A deli charges $3.45 for a pound of turkey. If Tim wants to purchase 2.4 pounds, how much will it cost?

Original content Copyright © by Holt McDougal. Additions and changes to the original content are the responsibility of the instructor.

Holt McDougal Mathematics

LESSON 3-5

Practice C

Multiplying Decimals

Find each product.

1.
$$\begin{array}{r} 9.86 \\ \times\ 0.3 \\ \hline \end{array}$$

2.
$$\begin{array}{r} 12.01 \\ \times\ 0.46 \\ \hline \end{array}$$

3.
$$\begin{array}{r} 7.05 \\ \times\ 0.03 \\ \hline \end{array}$$

4. $11.65 \cdot 0.23$

5. $24.54 \cdot 0.037$

6. $11.405 \cdot 2.91$

7. $0.058 \cdot 0.129$

8. $29.864 \cdot 5.13$

9. $100.86 \cdot 0.004$

Evaluate 17x for each value of x.

10. $x = 1.9$

11. $x = 0.005$

12. $x = 6.307$

13. $x = 11.215$

14. $x = 2.059$

15. $x = 75.844$

Evaluate.

16. $2.97n$ for $n = 1.8$

17. $12^2 + 1.9c$ for $c = 3.7$

18. $7^3 - 2x$ for $x = 0.54$

19. $1.6t + 3.056$ for $t = 2.09$

20. One year on Mercury is equal to 87.97 Earth days. One year on Pluto is three times the length of one Mercury year minus 16.21 days. How long is one year on Pluto?

21. One year on Earth is equal to 365.30 days. One year on Mars is twice the length of one Earth year minus 43.6 days. How long is one year on Mars?

Original content Copyright © by Holt McDougal. Additions and changes to the original content are the responsibility of the instructor.

Holt McDougal Mathematics

LESSON 3-5

Review for Mastery
Multiplying Decimals

You can use a model to help you multiply a decimal by a whole number.

Find the product of 0.12 and 4, using a 10 by 10 grid.

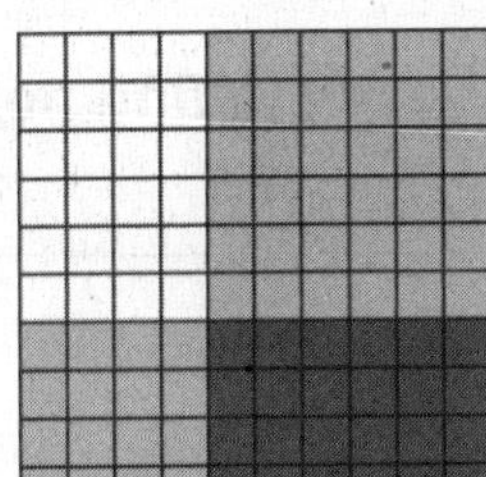

Shade 4 groups of 12 squares. Count the number of shaded squares. Since you have shaded 48 of the 100 squares, $0.12 \cdot 4 = 0.48$.

Find each product.

1. $0.23 \cdot 3$

2. $0.41 \cdot 2$

3. $0.011 \cdot 5$

4. $0.32 \cdot 2$

5. $0.15 \cdot 3$

6. $0.42 \cdot 2$

7. $0.04 \cdot 8$

8. $0.22 \cdot 4$

You can also use a model to help you multiply a decimal by a decimal.

Find the product of 0.4 and 0.6.

$0.4 \cdot 0.6 = 0.24$

Find each product.

9. $0.2 \cdot 0.8$

10. $0.7 \cdot 0.9$

11. $0.5 \cdot 0.5$

12. $0.3 \cdot 0.6$

13. $0.5 \cdot 0.2$

14. $0.4 \cdot 0.4$

15. $0.1 \cdot 0.9$

16. $0.4 \cdot 0.7$

Original content Copyright © by Holt McDougal. Additions and changes to the original content are the responsibility of the instructor.

Holt McDougal Mathematics

Challenge
Decimal Growth

Use the growth rate for each plant below to find how much it will grow in 1 week.

Eucalyptus Tree

Growth Rate:
2.5 cm per day

____

Bristlecone Pine Tree

Growth Rate:
0.009 mm per day

____

Trumpet Tree

Growth Rate:
0.28 in. per day

____

Use the growth rate for each plant below to find how much it will grow in 0.25 day.

Oak Tree

Growth Rate:
1.4 mm per day

____

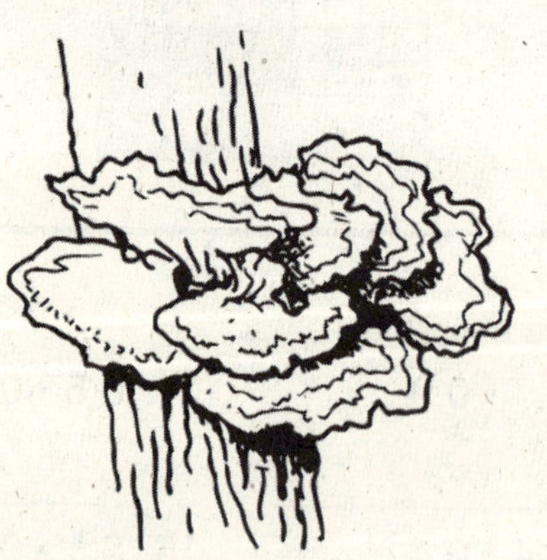

Lichens

Growth Rate:
0.0025 mm per day

____

Poplar Tree

Growth Rate:
0.118 in. per day

____

Original content Copyright © by Holt McDougal. Additions and changes to the original content are the responsibility of the instructor.

Holt McDougal Mathematics

LESSON 3-5

Problem Solving

Multiplying Decimals

Use the table to answer the questions.

United States Minimum Wage

Year	Hourly Rate
1940	$0.30
1950	$0.75
1960	$1.00
1970	$1.60
1980	$3.10
1990	$3.80
2000	$5.15

1. At the minimum wage, how much did a person earn for a 40-hour workweek in 1950?

   ____

2. At the minimum wage, how much did a person earn for working 25 hours in 1970?

   ____

3. If you had a minimum-wage job in 1990, and worked 15 hours a week, how much would you have earned each week?

   ____

4. About how many times higher was the minimum wage in 1960 than in 1940?

   ____

Circle the letter for the correct answer.

5. Ted's grandfather had a minimum-wage job in 1940. He worked 40 hours a week for the entire year. How much did Ted's grandfather earn in 1940?

   A $12.00

   B $624.00

   C $642.00

   D $6,240.00

6. Marci's mother had a minimum-wage job in 1980. She worked 12 hours a week. How much did Marci's mother earn each week?

   F $3.72

   G $37.00

   H $37.10

   J $37.20

7. Having one dollar in 1960 is equivalent to having $5.82 today. If you worked 40 hours a week in 1960 at minimum wage, how much would your weekly earnings be worth today?

   A $40.00

   B $5.82

   C $232.80

   D $2,328.00

8. In 2000, Cindy had a part-time job at a florist, where she earned minimum wage. She worked 18 hours each week for the whole year. How much did she earn from this job in 2000?

   F $927.00

   G $4,820.40

   H $10,712.00

   J $2,142.40

Original content Copyright © by Holt McDougal. Additions and changes to the original content are the responsibility of the instructor.

Holt McDougal Mathematics

LESSON 3-5

Reading Strategies
Use a Visual Tool

Each grid shows 0.15 shaded.

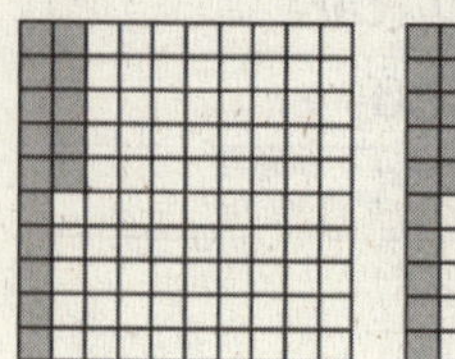 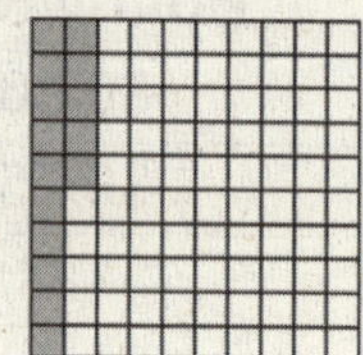 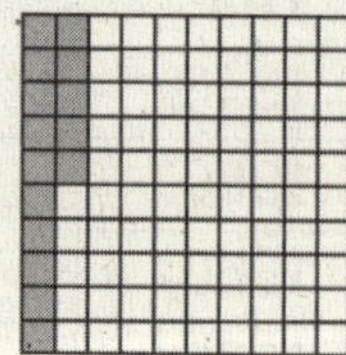

You can add the decimals to → 0.15 + 0.15 + 0.15 = 0.45
find how much of the grids
are shaded.

You can multiply 0.15 by 3. →

$$\begin{array}{r} \overset{1}{0.15} \\ \times\ \ 3 \\ \hline 0.45 \end{array}$$

Use these grids to complete the problems below.

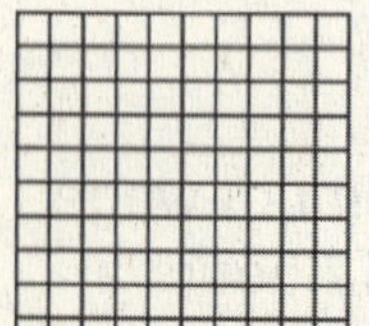  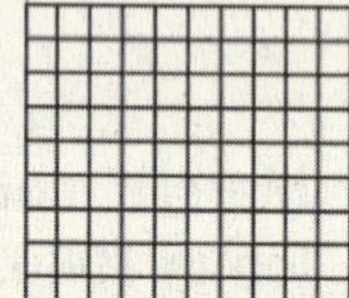 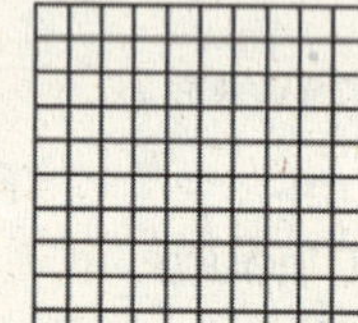

1. Shade 0.23 in each of the 4 grids.
2. Write an addition problem for the shaded grids.

____

3. Find the sum of your addition problem.

____

4. Write a multiplication problem for your shaded picture.

____

5. Find the product of your multiplication problem.

____

Original content Copyright © by Holt McDougal. Additions and changes to the original content are the responsibility of the instructor.

Puzzles, Twisters & Teasers

LESSON 3-5

Maria's Twice Cut Cake

Today is Maria's birthday. Maria's coworkers, Bob, Makiko, José, and Edna, bought her a cake. Unfortunately, Maria and her coworkers are dieting, so none of them wanted a whole piece of cake. They each wanted a fraction of a piece. Maria wanted 0.2 of a piece, Bob wanted 0.16 of a piece, Makiko wanted 0.32 of a piece, José wanted 0.225 of a piece and Edna wanted just 0.095 of a piece. Because Maria was not a mathematician, she misunderstood her coworkers—she thought that they wanted the entire cake divided into 5 pieces. So she cut the cake as shown.

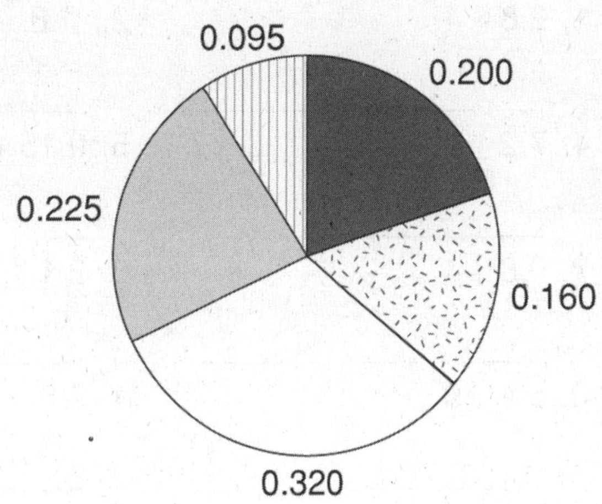

When the cake was cut, the coworkers received the wrong pieces. Maria got Bob's piece, Bob got Makiko's piece, Makiko got José's piece, José got Edna's piece, and Edna got Maria's piece.

Once they received their pieces, each coworker assumed they received a full piece of cake so each of them cut off the amount they had originally requested.

For example, Maria received Bob's 0.16 of the cake—so she ate her 0.2 piece of that. Mathematically she ate 0.2 x 0.16 = 0.032 piece of the cake. The question is: Who ate the most cake? List how much of the entire cake each of the five coworkers ate, then decide who ate the most cake.

Maria has 0.032 of a piece.

Bob has ________________

Makiko has ________________

José has ________________

Edna has ________________

So, who ate the most cake? ________________

Original content Copyright © by Holt McDougal. Additions and changes to the original content are the responsibility of the instructor.

                                        Holt McDougal Mathematics

LESSON 3-6

Practice A

Dividing Decimals by Whole Numbers

Find each quotient.

1. $2.8 \div 4$

2. $1.8 \div 2$

3. $3.6 \div 6$

4. $7.2 \div 9$

5. $0.15 \div 3$

6. $4.8 \div 8$

7. $0.8 \div 4$

8. $2.1 \div 7$

9. $0.32 \div 4$

10. $5.4 \div 9$

11. $3.5 \div 5$

12. $0.2 \div 2$

Evaluate $2.4 \div x$ for each given value of x.

13. $x = 8$

14. $x = 2$

15. $x = 3$

16. $x = 4$

17. $x = 6$

18. $x = 12$

19. A six-pack of orange soda costs $4.20. How much does each can in the pack cost?

20. It rained 2.7 inches in July and 2.1 inches in August. What was the average rainfall for those two months?

Original content Copyright © by Holt McDougal. Additions and changes to the original content are the responsibility of the instructor.

Holt McDougal Mathematics

LESSON 3-6

Practice B

Dividing Decimals by Whole Numbers

Find each quotient.

1. $0.81 \div 9$

2. $1.84 \div 4$

3. $7.2 \div 6$

4. $13.6 \div 8$

5. $4.55 \div 5$

6. $29.6 \div 8$

7. $15.57 \div 9$

8. $0.144 \div 12$

9. $97.5 \div 3$

10. $0.0025 \div 5$

11. $2.84 \div 8$

12. $18.9 \div 3$

Evaluate $2.094 \div x$ for each given value of x.

13. $x = 2$

14. $x = 4$

15. $x = 12$

16. $x = 20$

17. $x = 15$

18. $x = 30$

19. There are three grizzly bears in the city zoo. Yogi weighs 400.5 pounds, Winnie weighs 560.35 pounds, and Nyla weighs 618.29 pounds. What is the average weight of the three bears?

20. The bill for dinner came to $75.48. The four friends decided to leave a $15.00 tip. If they shared the bill equally, how much will they each pay?

Original content Copyright © by Holt McDougal. Additions and changes to the original content are the responsibility of the instructor.

Holt McDougal Mathematics

LESSON 3-6

Practice C
Dividing Decimals by Whole Numbers

Find each quotient.

1. $2.36 \div 8$

2. $0.1488 \div 3$

3. $72.654 \div 6$

4. $8.523 \div 9$

5. $115.8 \div 12$

6. $0.952 \div 17$

7. $46.545 \div 29$

8. $14.795 \div 55$

9. $0.2808 \div 75$

Evaluate $x \div 6$ for each value of x.

10. $x = 4.8$

11. $x = 0.54$

12. $x = 0.024$

13. $x = 1.08$

14. $x = 0.42$

15. $x = 0.0012$

Evaluate.

16. $n \div 19$ for $n = 28.5$

17. $(6^2 + 1.35) \div c$ for $c = 5$

18. $4^3 - (0.81 \div x)$ for $x = 9$

19. $3.5t \div 4$ for $t = 19.36$

20. As of 2000, there were 281.42 million people in the United States. If the same number of people lived in each of the 50 states, what would have been the population of each state in 2000?

21. In a gymnastics competition, Kim scored 9.4, 9.7, 9.9, and 9.8. Tamara scored 9.5, 9.2, 9.7, and 9.6. Who had the highest average score?

Original content Copyright © by Holt McDougal. Additions and changes to the original content are the responsibility of the instructor.

Holt McDougal Mathematics

LESSON 3-6
Review for Mastery
Dividing Decimals by Whole Numbers

You can use decimal grids to help you divide decimals by whole numbers.

To divide 0.35 by 7, first shade in a decimal grid to show thirty-five hundredths.

0.35 ÷ 7 means "divide 0.35 into 7 equal groups." Show this on the decimal grid.

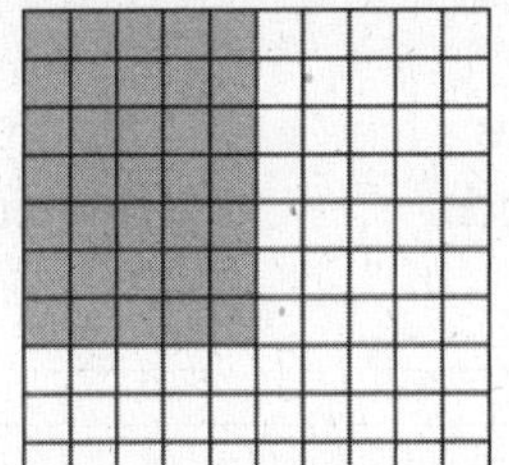

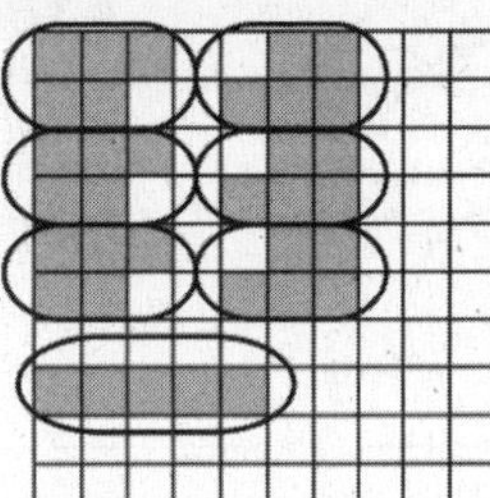

The number of units in each group is the quotient.

So, 0.35 ÷ 7 = 0.05.

Use decimal grids to find each quotient.

1. 0.24 ÷ 4

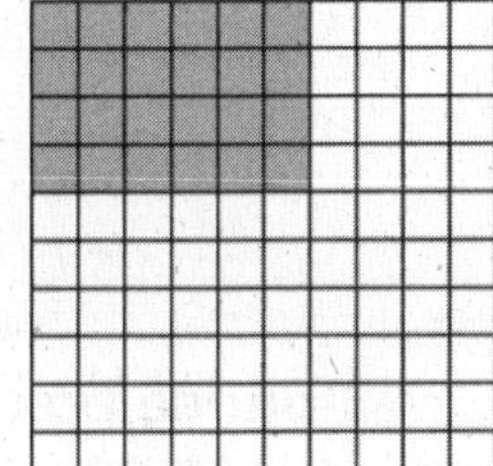

2. 0.48 ÷ 12

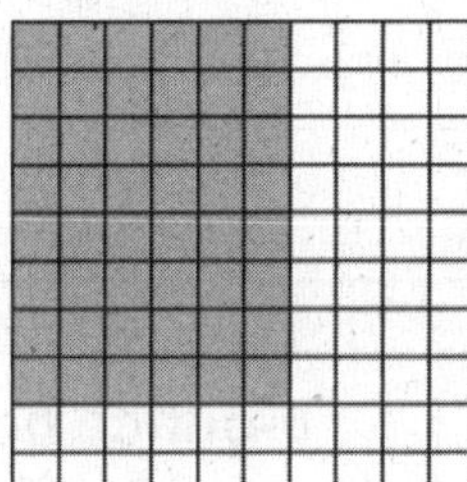

3. 0.50 ÷ 10

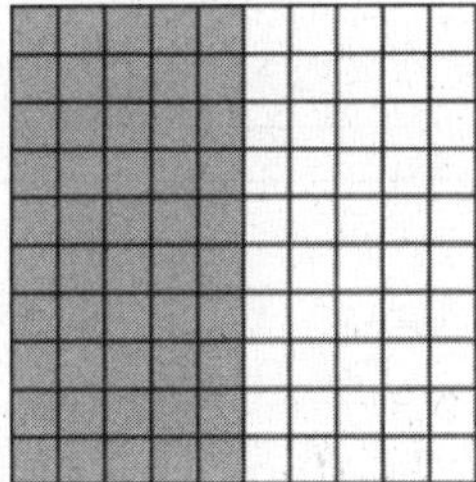

4. 0.98 ÷ 7

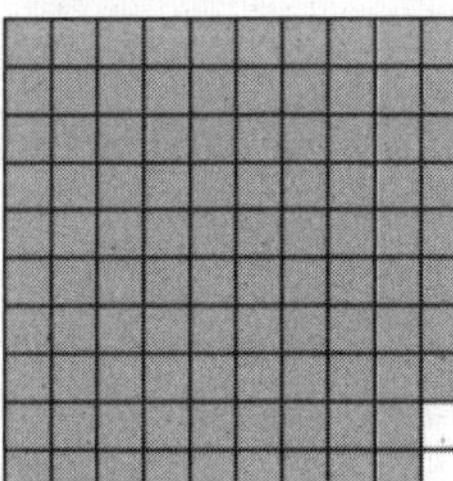

5. 0.6 ÷ 5   6. 0.78 ÷ 6

7. 0.99 ÷ 11   8. 0.32 ÷ 4

_____________   _____________   _____________   _____________

Original content Copyright © by Holt McDougal. Additions and changes to the original content are the responsibility of the instructor.

Holt McDougal Mathematics

LESSON 3-6

Challenge

Get the Best Deal

Grocery stores often sell items in different quantities, package sizes, and unit prices. A unit price is the price for one unit of an item. To get the best deal, you should buy each item with the lowest unit price. Find each unit price and determine the best deal.

	1 for $0.69	6 for $2.70	12 for $4.80
	Unit price (per pound) ________	Unit price (per pound) ________	Unit price (per pound) ________

Best deal: ________________________________

	1 pound for $0.75	2 pounds for $1.70	5 pounds for $4.05
	Unit price (per pound) ________	Unit price (per pound) ________	Unit price (per pound) ________

Best deal: ________________________________

	6-ounce box for $1.98	12-ounce box for $3.72	16-ounce box for $5.28
	Unit price (per ounce) ________	Unit price (per ounce) ________	Unit price (per ounce) ________

Best deal: ________________________________

	6-pack for $1.08	12-pack for $2.64	24-pack for $4.08
	Unit price (per can) ________	Unit price (per can) ________	Unit price (per can) ________

Best deal: ________________________________

Original content Copyright © by Holt McDougal. Additions and changes to the original content are the responsibility of the instructor.

Holt McDougal Mathematics

LESSON 3-6	**Problem Solving**

Dividing Decimals by Whole Numbers

Write the correct answer.

1. Four friends had lunch together. The total bill for lunch came to $33.40, including tip. If they shared the bill equally, how much did they each pay?

2. There are 7.2 milligrams of iron in a dozen eggs. Because there are 12 eggs in a dozen, how many milligrams of iron are in 1 egg?

3. Kyle bought a sheet of lumber 8.7 feet long to build fence rails. He cut the strip into 3 equal pieces. How long is each piece?

4. An albatross has a wingspan greater than the length of a car—3.7 meters! Wingspan is the length from the tip of one wing to the tip of the other wing. What is the length of each albatross wing (assuming wing goes from center of body)?

Circle the letter of the correct answer.

5. The City Zoo feeds its three giant pandas 181.5 pounds of bamboo shoots every day. Each panda is fed the same amount of bamboo. How many pounds of bamboo does each panda eat every day?

   A 6.05 pounds

   B 60.5 pounds

   C 61.5 pounds

   D 605 pounds

6. Emma bought 22.5 yards of cloth to make curtains for two windows in her apartment. She used the same amount of cloth on each window. How much cloth did she use to make each set of curtains?

   F 1.125 yards

   G 10.25 yards

   H 11.25 yards

   J 11.52 yards

7. Aerobics classes cost $153.86 for 14 sessions. What is the fee for one session?

   A $10.99

   B $1.99

   C about $25.00

   D about $20.00

8. An entire apple pie has 36.8 grams of saturated fat. If the pie is cut into 8 slices, how many grams of saturated fat are in each slice?

   F 4.1 grams

   G 0.46 grams

   H 4.6 grams

   J 4.11 grams

Original content Copyright © by Holt McDougal. Additions and changes to the original content are the responsibility of the instructor.

Holt McDougal Mathematics

Reading Strategies

LESSON 3-6

Use a Visual Tool

You can use a hundred grid to show division with decimals.

The grid shows 0.15. →

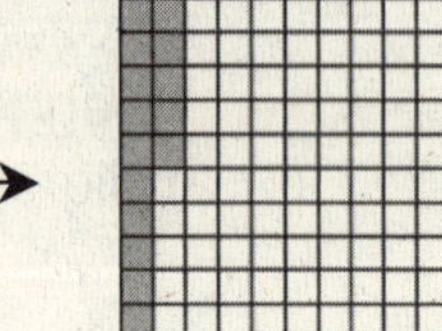

0.15 ÷ 3 means "separate
0.15 into 3 equal groups." →

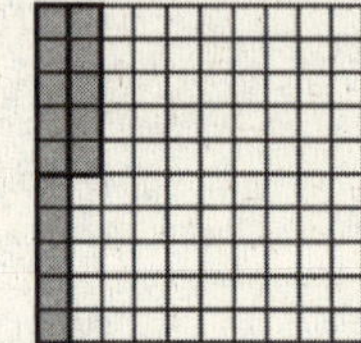

0.15 ÷ 3 makes 3 equal
groups of 0.05. →

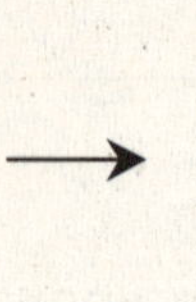 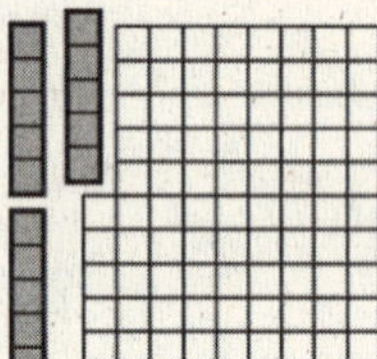

0.15 ÷ 3 = 0.05

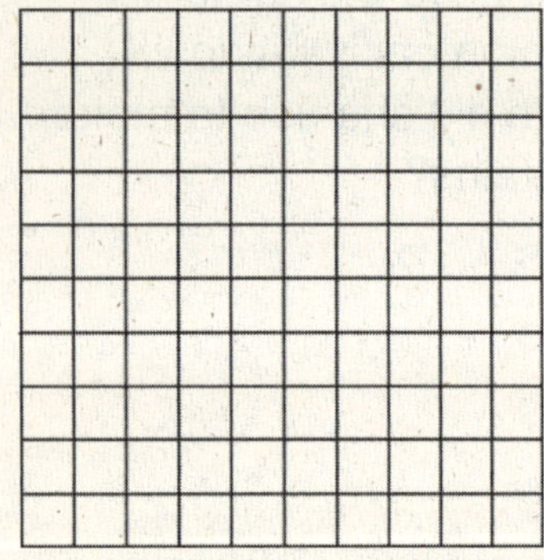

Use the grid to complete Exercises 1–4.

1. Shade 0.60 of the grid.

2. Divide the grid into 3 equal groups.

3. Write the decimal amount in each of the 3 groups. _______________

4. Write a division problem for the picture you have created.

____

Original content Copyright © by Holt McDougal. Additions and changes to the original content are the responsibility of the instructor.

Holt McDougal Mathematics

LESSON
3-6

Puzzles, Twisters & Teasers

One Nation, Indivisible. . .

What state had the first traffic light?

We know the United States consists of fifty individual states. Often
states want to make improvements to roads, state parks, or schools.
They look to the federal government for monies to pay for those
improvements.

For each of the improvements listed below, the government has
decided which states will receive money. Divide the money for each
item by the number of states receiving the money to find how much
money each state gets.

ITEM	MONEY	RECIPIENT STATES	MONEY PER STATE
State Parks	$193.45 per acre	CT, CO, WY, UT, AL	
Farming	$639.24 per acre	NE, KS, MO	
Roads	$534.36 per mile	WV, GA, MS, NC, OR, WA	
Security	$94.56 per government building	MD, PA	
Schools	$143.04 per classroom	LA, OK, ME, RI, KY, AL, WI, HI	
Science/ Research	$193.20 per laboratory	FL, TX, MA, CA, AZ	
Museums	$884.79 per display	NY, NM, NH	

To answer the riddle, find the column labeled with largest amount of
money and the row labeled with the smallest amount of money.

	$305.17	$296.40	$294.93	$213.08
$22.04	Maryland	North Dakota	Oklahoma	Oregon
$14.45	Ohio	Florida	Texas	Vermont
$17.88	Iowa	Minnesota	New York	Louisiana
$38.69	Pennsylvania	Arkansas	Maine	Arizona

So, what state do you think had the first
traffic light?

____

Original content Copyright © by Holt McDougal. Additions and changes to the original content are the responsibility of the instructor.

Holt McDougal Mathematics

LESSON 3-7

Practice A

Dividing by Decimals

Find each quotient.

1. $2.4 \div 0.4$

2. $1.4 \div 0.2$

3. $4.8 \div 0.6$

4. $8.1 \div 0.9$

5. $1.8 \div 0.3$

6. $6.4 \div 0.8$

7. $3.3 \div 0.3$

8. $2.6 \div 1.3$

9. $7.2 \div 1.2$

10. $7.5 \div 1.5$

11. $6.0 \div 0.5$

12. $9.9 \div 1.1$

Evaluate $4.8 \div x$ for each value of x.

13. $x = 0.2$

14. $x = 0.4$

15. $x = 0.3$

16. $x = 0.6$

17. $x = 0.8$

18. $x = 1.2$

19. Antonio spent $5.60 on cashews. They cost $1.40 per pound. How many pounds of cashews did Antonio buy?

20. Over several months, a scientist measured a total of 6.3 inches of snow. The average snowfall each month was 2.1 inches. How many months did the scientist measure the snow?

Original content Copyright © by Holt McDougal. Additions and changes to the original content are the responsibility of the instructor.

LESSON 3-7	**Practice B**

Practice B
Dividing by Decimals

Find each quotient.

1. $9.0 \div 0.9$
2. $29.6 \div 3.7$
3. $10.81 \div 2.3$

4. $10.5 \div 1.5$
5. $15.36 \div 4.8$
6. $9.75 \div 1.3$

7. $20.4 \div 5.1$
8. $37.5 \div 2.5$
9. $9.24 \div 1.1$

10. $16.56 \div 6.9$
11. $28.9 \div 8.5$
12. $14.35 \div 0.7$

Evaluate $x \div 1.2$ for each value of x.

13. $x = 40.8$
14. $x = 1.8$
15. $x = 10.8$

16. $x = 14.4$
17. $x = 4.32$
18. $x = 0.06$

19. Anna is saving $6.35 a week to buy a computer game that costs $57.15. How many weeks will she have to save to buy the game?

20. Ben ran a 19.5-mile race last Saturday. His average speed during the race was 7.8 miles per hour. How long did it take Ben to finish the race?

Original content Copyright © by Holt McDougal. Additions and changes to the original content are the responsibility of the instructor.

Holt McDougal Mathematics

LESSON 3-7

Practice C

Dividing by Decimals

Find each quotient.

1. $4.75 \div 2.5$

2. $34.04 \div 4.6$

3. $10.0 \div 1.25$

4. $283.62 \div 8.7$

5. $168.75 \div 6.75$

6. $0.1092 \div 0.013$

7. $7.7293 \div 3.7$

8. $97.206 \div 5.1$

9. $0.489807 \div 0.081$

Evaluate $15.65 \div x$ for each value of x.

10. $x = 0.2$

11. $x = 0.4$

12. $x = 0.5$

13. $x = 0.8$

14. $x = 1.6$

15. $x = 2.5$

Evaluate.

16. $n \div 7.8$ for $n = 26.988$

17. $(7^2 - 32.9) \div c$ for $c = 3.5$

18. $18.67 - (0.216 \div x)$ for $x = 0.02$

19. $4.4t \div 1.6$ for $t = 16.92$

20. The sum of two decimal numbers is 3.9. Their difference is 0.9, and their product is 3.6. What are the two numbers?

21. The sum of two decimal numbers is 5.3. Their difference is 1.7, and their product is 6.3. What are the two numbers?

Original content Copyright © by Holt McDougal. Additions and changes to the original content are the responsibility of the instructor.

Holt McDougal Mathematics

LESSON 3-7	# Review for Mastery
	## *Dividing by Decimals*

You can use powers of ten to help you divide a decimal by a decimal.

To divide 0.048 by 0.12, first multiply each number by the least power of ten that makes the divisor a whole number.

$0.048 \div 0.12$

$0.12 \cdot 10^2 = 12$ Move the decimal point 2 places to the right.

$0.048 \cdot 10^2 = 4.8$ Move the decimal point 2 places to the right.

Then divide.

$4.8 \div 12$ **Step 1:** Divide as you would divide a whole number by a whole number.

$$\begin{array}{r} 0.4 \\ 12\overline{)4.8} \\ \underline{4\,8} \\ 0 \end{array}$$

Step 2: Think $48 \div 12 = 4$.

Step 3: Bring the decimal into the quotient and add a zero placeholder if necessary.

So, $0.048 \div 0.12 = 0.4$.

Find each quotient.

1. $0.7\overline{)0.42}$ 2. $0.08\overline{)0.4}$ 3. $0.5\overline{)0.125}$ 4. $0.02\overline{)0.3}$

5. $0.4\overline{)0.08}$ 6. $0.9\overline{)0.63}$ 7. $0.008\overline{)0.4}$ 8. $0.04\overline{)0.032}$

9. $0.3\overline{)0.06}$ 10. $0.04\overline{)0.2}$ 11. $0.007\overline{)4.9}$ 12. $0.6\overline{)0.012}$

Original content Copyright © by Holt McDougal. Additions and changes to the original content are the responsibility of the instructor.

Holt McDougal Mathematics

<table><tr><td>LESSON
3-7</td><td>

Challenge
Cutting Decimals

</td></tr></table>

The strips of cloth below need to be cut into equal pieces of given lengths. Draw lines on each strip of cloth to show how many pieces will be cut.

1. **Total Length: 9.8 yards   Piece Length: 1.4 yards**

2. **Total Length: 2.5 yards   Piece Length: 0.5 yards**

3. **Total Length: 10.2 yards   Piece Length: 1.7 yards**

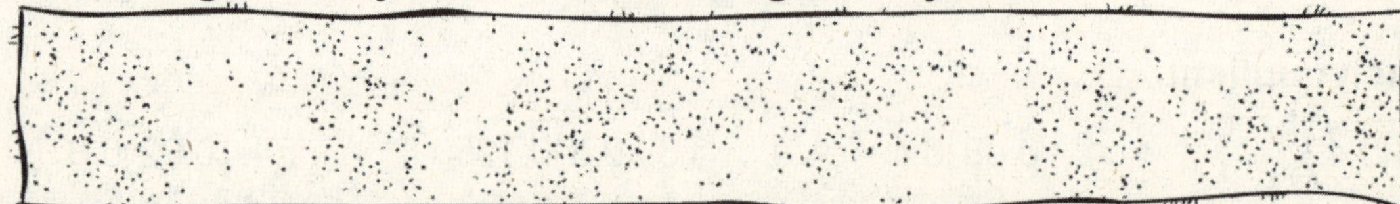

4. **Total Length: 6.4 yards   Piece Length: 0.8 yards**

5. **Total Length: 13.6 yards   Piece Length: 3.4 yards**

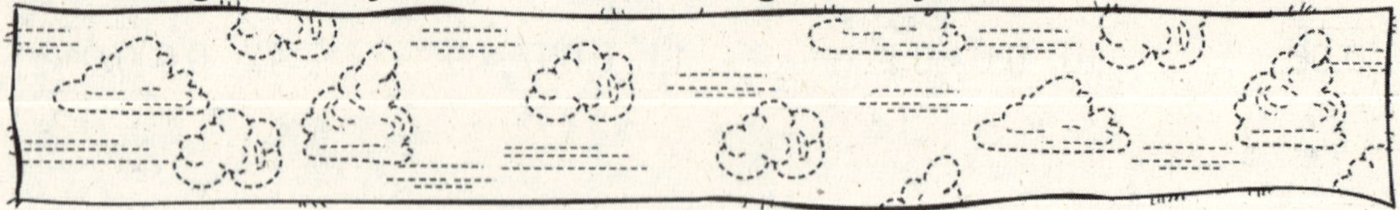

Original content Copyright © by Holt McDougal. Additions and changes to the original content are the responsibility of the instructor.

<table>
<tr><td>LESSON
3-7</td><td>

Problem Solving
Dividing by Decimals
</td></tr>
</table>

Write the correct answer.

1. Peter spent $6.75 on wire to build a rabbit hutch. Wire costs $0.45 per foot. How many feet of wire did Peter buy?

2. Jamal drove 195.3 miles in 3.5 hours. On average, how many miles per hour did he drive?

3. Lisa's family drove 830.76 miles to visit her grandparents. Lisa calculated that they used 30.1 gallons of gas. How many miles per gallon did the car average?

4. A chef bought 84.5 pounds of ground beef. She uses 0.5 pound of ground beef for each hamburger. How many hamburgers can she make?

Circle the letter of the correct answer.

5. Mark earned $276.36 for working 23.5 hours last week. He earned the same amount of money for each hour that he worked. What is Mark's hourly rate of pay?

   A $1.17

   B $10.76

   C $11.76

   D $117.60

6. Alicia wants to cover a section of her wall that is 2 feet wide and 12 feet long with mirrors. Each mirror tile is 2 feet wide and 1.5 feet long. How many mirror tiles does she need to cover that section?

   F 4 tiles

   G 6 tiles

   H 8 tiles

   J 12 tiles

7. John ran the city marathon in 196.5 minutes. The marathon is 26.2 miles long. On average, how many miles per hour did John run the race?

   A 7 miles per hour

   B 6.2 miles per hour

   C 8 miles per hour

   D 8.5 miles per hour

8. Shaneeka is saving $5.75 of her allowance each week to buy a new camera that costs $51.75. How many weeks will she have to save to have enough money to buy it?

   F 9 weeks

   G 9.5 weeks

   H 8.1 weeks

   J 8 weeks

Original content Copyright © by Holt McDougal. Additions and changes to the original content are the responsibility of the instructor.

Holt McDougal Mathematics

LESSON 3-7

Reading Strategies
Make Predictions

Study the examples below. Look for patterns in the divisor and quotient.

Dividend		Divisor		Quotient
400	÷	20	=	20
400	÷	2	=	200
400	÷	0.2	=	2,000
400	÷	0.02	=	20,000

As the divisor is divided by 10, the quotient is multiplied by 10.

Use the information above to answer Exercises 1–3.

1. Predict the divisor for the next problem in this pattern.

   ____

2. Predict the quotient for the next problem in this pattern.

   ____

3. Write the next division problem and quotient for this pattern.

   ____

Study the pattern created by these division problems. Use the pattern to answer Exercises 4–6.

Dividend		Divisor		Quotient
900	÷	30	=	30
900	÷	3	=	300
900	÷	0.3	=	3,000

4. Predict the next divisor in this pattern.

   ____

5. Predict the next quotient in this pattern.

   ____

6. Write the division problem and quotient that you predict would come next.

   ____

Original content Copyright © by Holt McDougal. Additions and changes to the original content are the responsibility of the instructor.

Holt McDougal Mathematics

<table>
<tr><td>LESSON
3-7</td><td></td></tr>
</table>

Puzzles, Twisters & Teasers
Bargain Hunter

You are a professional shopper. You've been asked to use your shopping skills to find the best buy among the following cereals. You must determine which of these cereals costs the least per ounce.

Type of Cereal	Size of Box	Price per Box
Frostee O's	22.5 oz	$12.60
Super Sugar Loops	9.9 oz	$2.97
Marshmallow Bonanza	22.8 oz	$3.42
Tyrannosaurus Rings	6.6 oz	$3.30
Wheat and Rice Explosion	35.6 oz	$3.56
Giant Frosted Oatees	10.0 oz	$6.10
Ned's Enormous O's of Sugar	15.5 oz	$1.86

Type of Cereal	Price per Oz
Frostee O's	
Super Sugar Loops	
Marshmallow Bonanza	
Tyrannosaurus Rings	
Wheat and Rice Explosion	
Giant Frosted Oatees	
Ned's Enormous O's of Sugar	

Do you know what lies at the bottom of the ocean and twitches?

To find out, fill in the capital letters from the *second* best bargain cereal in order in the first row of spaces below. Then fill in the capital letters in order from the *best* bargain cereal and you should have the answer!

A ___ ___ ___ R V ___ U ___
___ ___ ___ C K

Original content Copyright © by Holt McDougal. Additions and changes to the original content are the responsibility of the instructor.

Holt McDougal Mathematics

<table>
<tr><td>LESSON
3-8</td><td></td></tr>
</table>

Practice A
Interpreting the Quotient

Circle the letter of the correct answer.

1. Hamburger rolls come in packs of 8. How many packs should you buy to have 60 rolls?

   A 8

   B 6

   C 5

   D 7

2. Each pack of hamburger rolls costs $1.50. How many packs can you buy with $8.00?

   F 6

   G 5

   H 4

   J 8

3. How many 0.6-pound hamburgers can you make with 7.8 pounds of ground beef?

   A 13

   B 14

   C 10

   D 16

4. You spend a total of $5.10 for 3 pounds of ground beef. How much does the ground beef cost per pound?

   F $0.70

   G $0.17

   H $15.30

   J $1.70

Write the correct answer.

5. Four friends equally shared the cost of buying supplies for the class picnic. The supplies cost a total of $12.40. How much did they each pay?

   ____

6. In all, 20 people are going to the picnic. Each van seats 6 people. How many vans are needed to take everyone to the picnic?

   ____

7. Plastic forks come in packs of 6. If you need 40 forks for the picnic, how many packs should you buy?

   ____

8. You spent a total of $9.60 on paper plates for the picnic. Each pack costs $1.20. How many packs of paper plates did you buy?

   ____

Original content Copyright © by Holt McDougal. Additions and changes to the original content are the responsibility of the instructor.

   Holt McDougal Mathematics

LESSON
3-8

Practice B

Interpreting the Quotient

Circle the letter of the correct answer.

1. You spent a total of $6.75 for 15 yards of ribbon. How much did the ribbon cost per yard?

   A $0.50

   B $0.45

   C $1.35

   D $1.45

2. Buttons come in packs of 12. How many packs should you buy if you need 100 buttons?

   F 10

   G 8

   H 9

   J 12

3. Your sewing cabinet has compartments that hold 8 spools of thread each. You have 50 spools of thread. How many compartments can you fill?

   A 6

   B 7

   C 5

   D 8

4. You spent a total of $35.75 for velvet cloth. Each yard of the velvet costs $3.25. How many yards did you buy?

   F 10

   G 10.5

   H 11

   J 11.5

Write the correct answer.

5. You used a total of 67.5 yards of cotton material to make costumes for the play. Each costume used 11.25 yards of cloth. How many costumes did you make?

   ____

6. You are saving $17.00 each week to buy a new sewing machine that costs $175.50. How many weeks will you have to save to have enough money to buy the sewing machine?

   ____

7. Sequins come in packs of 75. You use 12 sequins on each costume. If you have one pack of sequins, how many costumes can you make?

   ____

8. You pay $26.28 for a subscription to Sewing Magazine. You get an issue every month for a year. How much does each issue cost?

   ____

Original content Copyright © by Holt McDougal. Additions and changes to the original content are the responsibility of the instructor.

Holt McDougal Mathematics

LESSON
3-8

Practice C

Interpreting the Quotient

Write the correct answer.

1. You live in Detroit, Michigan. You and your parents will be driving to Cincinnati, Ohio, to visit your grandparents. The trip is a total of 264.1 miles. Your family car averages 27.8 miles per gallon. How many gallons of gas will you use on the trip?

2. You plan on taking 100 photographs during your trip. You want to choose one kind of film to buy, either a 24-photo roll, or an 18-photo roll. How many packs of each would you need to buy?

3. The gas tank of your parents' car holds 12.9 gallons of gas. Because their car averages 27.8 miles per gallon, how many times will they have to fill up the gas tank to drive to Cincinnati and back?

4. The first time you stop at a gas station during your trip, your parents spend $13.80 for 11.5 gallons of gas. How much does the gas cost per gallon?

5. You stop for lunch during the trip and spend a total of $11.38. Your cheeseburger costs $3.25. Both of your parents have a slice of pizza that each cost $2.19. You all get the same drink. How much did each drink cost?

6. During the trip your parents drive 211.05 miles on different highways. You calculate that you spent 3.5 hours driving on highways. What was the average speed your parents drove during that part of the trip?

7. You saved for 16 weeks before your trip to buy a present for your grandparents that cost $67.36. How much did you save each week?

8. Your parents budgeted $20 a month for long-distance calls. It costs $0.37 per minute for long distance calls. How many minutes can you talk to your grandparents each month?

Original content Copyright © by Holt McDougal. Additions and changes to the original content are the responsibility of the instructor.

Holt McDougal Mathematics

<table>
<tr><td>LESSON
3-8</td><td></td></tr>
</table>

Review for Mastery

Interpreting the Quotient

There are three ways the decimal part of a quotient can be interpreted when you solve a problem.

If the question asks for an exact number, use the entire quotient.

If the question asks how many whole groups are needed to put the dividend into a group, round the quotient up to the next whole number.

To interpret the quotient, decide what the question is asking.

In the school library, there are tables that seat 4 students each. If there are 30 students in a class, how many tables are needed to seat all of the students?

To solve, divide 30 by 4.

$30 \div 4 = 7.5$

The question is asking how many tables (whole groups) are needed to put all of the students in the class (dividend) into a group.

So, round 7.5 up to the next whole number.

8 tables are needed to seat all of the students.

Interpret the quotient to solve each problem.

1. A recipe that serves 6 requires 9 cups of milk. How much milk is needed for each serving?

2. A storage case holds 24 model cars. Marla has 84 model cars. How many storage cases does she need to store all of her cars?

3. Kenny has $4.25 to spend at the school carnival. If game tickets are $0.50 each, how many games can Kenny play?

Original content Copyright © by Holt McDougal. Additions and changes to the original content are the responsibility of the instructor.

Holt McDougal Mathematics

<table><tr><td>LESSON
3-8</td></tr></table>

Challenge

Plan a Party!

**You are in charge of buying supplies for the class party.
There are 30 students in your class. Use the party supply
store advertisement below to plan what to buy. After you pay
for all the items, the total cost will be divided evenly among all
the students.**

Shopping List

Item	Number of Items You Want Per Person	Number of Packs to Buy	Number of Left Over Items	Total Price of Items
Invitations	1			
Paper plates	1			
Plastic cups	2			
Paper napkins	2			
Plastic forks	1			
			Grand Total Price:	
			Cost Per Student:	

Original content Copyright © by Holt McDougal. Additions and changes to the original content are the responsibility of the instructor.

     Holt McDougal Mathematics

LESSON 3-8 — Problem Solving
Interpreting the Quotient

Write the correct answer.

1. Five friends split a pizza that costs $16.75. If they shared the bill equally, how much did they each pay?

2. There are 45 choir members going to the recital. Each van can carry 8 people. How many vans are needed?

3. Tara bought 150 beads. She needs 27 beads to make each necklace. How many necklaces can she make?

4. Cat food costs $2.85 for five cans. Ben only wants to buy one can. How much will it cost?

Circle the letter of the correct answer.

5. Tennis balls come in cans of 3. The coach needs 50 tennis balls for practice. How many cans should he order?

   A 16 cans

   B 17 cans

   C 18 cans

   D 20 cans

6. The rainfall for three months was 4.6 inches, 3.5 inches, and 4.2 inches. What was the average monthly rainfall during that time?

   F 41 inches

   G 12.3 inches

   H 4.3 inches

   J 4.1 inches

7. Tom has $15.86 to buy marbles that cost $1.25 each. He wants to know how many marbles he can buy. What should he do after he divides?

   A Drop the decimal part of the quotient when he divides.

   B Drop the decimal part of the dividend when he divides.

   C Round the quotient up to the next highest whole number to divide.

   D Use the entire quotient of his division as the answer.

8. Mei needs 135 hot dog rolls for the class picnic. The rolls come in packs of 10. She wants to know how many packs to buy. What should she do after she divides?

   F Drop the decimal part of the quotient when she divides.

   G Drop the decimal part of the dividend when she divides.

   H Round the quotient up to the next highest whole number.

   J Use the entire quotient of her division as the answer.

Original content Copyright © by Holt McDougal. Additions and changes to the original content are the responsibility of the instructor.

Holt McDougal Mathematics

Reading Strategies

LESSON 3-8

Use Context

How the decimal portion of the quotient in a division problem is used depends upon the situation.

Situation 1 74 students are going on a field trip in cars. Each car can carry 5 students. How many cars are needed?

Divide 74 by 5. $\longrightarrow$ $74 \div 5 = 14.8$ cars

Reasoning 14 cars will not be enough for all students. You need 15 cars. The quotient 14.8 needs to be rounded up to 15 in this situation.

Situation 2 How many 8 oz servings are in a 44 oz can of juice?

Divide 44 by 8. $\longrightarrow$ $44 \div 8 = 5.5$ servings

Reasoning There are 5 full 8 oz servings in the can. The 0.5 serving is not 8 ounces. The quotient 5.5 is rounded down to 5 in this situation.

Situation 3 4 boys mowed a lawn for $35. How much money should each boy receive to share the money equally?

Divide $35 by 4. $\longrightarrow$ $\$35 \div 4 = \$8.75$

Reasoning The exact quotient of $8.75 states what each boy should receive. The exact quotient of $8.75 makes sense.

Tell whether you would round the quotient up, round the quotient down, or leave the exact quotient for each. Write to explain your choice.

1. You need 8 inches of ribbon to make a bow. How many bows can you make with 50 inches of ribbon? $50 \div 8 = 6.25$

   ____

2. Each lunch table seats 10 children. There are 155 children in the cafeteria for each lunch period. How many tables are needed? $155 \div 10 = 15.5$

   ____

Original content Copyright © by Holt McDougal. Additions and changes to the original content are the responsibility of the instructor.

  Holt McDougal Mathematics

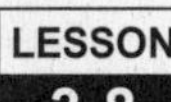

LESSON 3-8 — Puzzles, Twisters & Teasers
To Dine or Not to Dine?

John loves to have his friends over for dinner. He is concerned that he will not have enough food.

He knows that he has 8 cups of soup, 14 pounds of lasagna, 2.85 pounds of salad, and 7.3 pints of ice cream. John knows that, on average, each guest eats about 0.75 cups of soup, 1.2 pounds of lasagna, 0.3 pounds of salad, and 0.6 pint of ice cream. Use these numbers to help John decide how many people he can invite for dinner.

Soup _______________

Salad _______________

Lasagna _______________

Ice Cream _______________

7	T
8	O
9	U
10	M
11	L
12	N

Number of people John should invite: _______________

Do you know what John's gossiping friend had during dinner?

Find the letter that is paired with the number of people that John should invite and add it to the letter bank. Also, add one to the number of people John can invite, find the matching letter in the table above and add that letter to the letter bank as well.

Rearrange the letters in the letter bank to find the answer to the riddle.

He had a great sense of

____ ____ ____ ____ ____

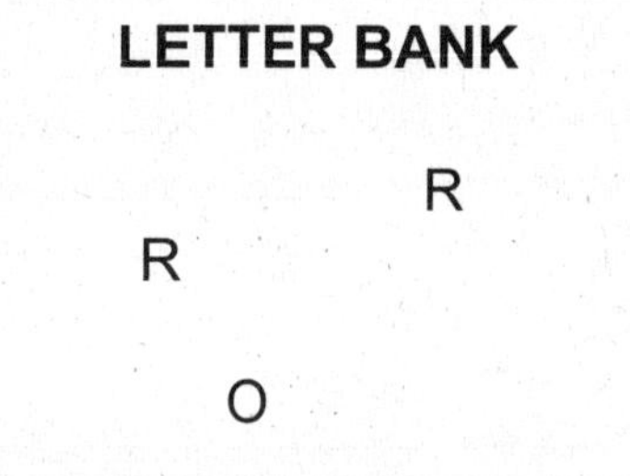

Original content Copyright © by Holt McDougal. Additions and changes to the original content are the responsibility of the instructor.

LESSON 3-9

Practice A

Solving Decimal Equations

Circle the letter of the correct answer.

1. If $7.2 + x = 9.7$, what is the value of x?

   A 16.9

   B 2.5

   C 2.9

   D 16.5

2. If $n \div 4 = 0.4$, what is the value of n?

   F 4.4

   G 0.1

   H 1.6

   J 10

Solve each equation. Check your answer.

3. $a - 0.4 = 1.3$

4. $2n = 1.8$

5. $0.8 + x = 1.3$

6. $p \div 4 = 0.7$

7. $w - 2.3 = 1.2$

8. $0.5q = 0.25$

9. $5.7 + s = 6.0$

10. $b \div 3 = 0.6$

11. $t - 3.1 = 1.6$

12. $3y = 1.5$

13. The length of a photograph is 2.1 inches, and the width is 3 inches. Solve the equation $a \div 2.1 = 3$ to find the area of the photograph.

14. It costs $0.90 to enlarge each photograph. If you want to enlarge 5 photos, how much will it cost in all?

Original content Copyright © by Holt McDougal. Additions and changes to the original content are the responsibility of the instructor.

Holt McDougal Mathematics

LESSON 3-9

Practice B

Solving Decimal Equations

Solve each equation. Check your answer.

1. $a - 2.7 = 4.8$

2. $b \div 7 = 1.9$

3. $w - 6.5 = 3.8$

4. $p \div 0.4 = 1.7$

5. $4.5 + x = 8$

6. $b \div 3 = 2.5$

7. $7.8 + s = 15.2$

8. $1.63q = 9.78$

9. $0.05 + x = 2.06$

10. $1.7n = 2.38$

11. $t - 6.08 = 12.59$

12. $9q = 16.2$

13. $w - 8.9 = 10.3$

14. $1.4n = 3.22$

15. $t - 12.7 = 0.8$

16. $3.8 + a = 6.5$

17. The distance around a square photograph is 12.8 centimeters. What is the length of each side of the photograph?

18. You buy two rolls of film for $3.75 each. You pay with a $10 bill. How much change should you get back?

Original content Copyright © by Holt McDougal. Additions and changes to the original content are the responsibility of the instructor.

Holt McDougal Mathematics

LESSON 3-9

Practice C
Solving Decimal Equations

Solve each equation. Check your answer.

1. $a - 0.089 = 12.5$

2. $b \div 2.8 = 4.7$

3. $w - 12.76 = 15.8$

4. $p \div 0.07 = 4.76$

5. $1.7806 + x = 2.009$

6. $b \div 6.5 = 9.7$

7. $8.3044 + s = 16.01$

8. $0.009q = 0.765$

9. $23.764 + x = 30.5$

10. $1.7n + 3.8 = 7.71$

11. $t - (8^2 + 0.36) = 0.5$

12. $9q + 0.2q = 27.6$

13. $8.4w - 0.67 = 15.29$

14. $1.4n + 4^3 = 85$

15. Andy bought 15 pounds of apples for $0.69 a pound, and 7.2 pounds of grapes for $3.65 a pound. He paid for the entire purchase with a $50 bill. How much change did he get back? He wants to make 25 pounds of fruit salad for the party. How much more fruit does he need?

16. Brenda's garden is 6.5 feet long and 8.76 feet wide. She uses half of the garden for vegetables. Then she equally divides the rest of the garden between flowers and herbs. How large is her vegetable garden? How much land does she use for growing flowers?

Original content Copyright © by Holt McDougal. Additions and changes to the original content are the responsibility of the instructor.

Holt McDougal Mathematics

LESSON 3-9

Review for Mastery

Solving Decimal Equations

You can write related equations for addition and subtraction equations.

$7.4 + 6.2 = 13.6$ $13.6 - 6.2 = 7.4$

Use related equations to solve each of the following.

A. $x + 4.5 = 7.9$

Think: $7.9 - 4.5 = x$

$x = 3.4$

Check $x + 4.5 = 7.9$

$3.4 + 4.5 \overset{?}{=} 7.9$ substitute

$7.9 = 7.9$

B. $x - 0.08 = 6.2$

Think: $6.2 + 0.08 = x$

$x = 6.28$

Check $x - 0.08 = 6.2$

$6.28 - 0.08 \overset{?}{=} 6.2$ substitute

$6.2 = 6.2$

Use related facts to solve each equation. Then check each answer.

1. $x + 8.7 = 12.9$

2. $x + 8.4 = 16.6$

3. $x - 2.65 = 7.8$

4. $x - 0.8 = 2.3$

You can write related equations for multiplication and division equations.

$3.2 \cdot 2.4 = 7.68$ $7.68 \div 2.4 = 3.2$

Use related equations to solve each of the following.

C. $3x = 1.5$

Think: $1.5 \div 3 = x$

$x = 0.5$

Check: $3x = 1.5$

$3 \cdot 0.5 \overset{?}{=} 1.5$ substitute

$1.5 = 1.5$

D. $x \div 6 = 1.2$

Think: $1.2 \cdot 6 = x$

$x = 7.2$

Check: $x \div 6 = 1.2$

$7.2 \div 6 \overset{?}{=} 1.2$ substitute

$1.2 = 1.2$

Use related facts to solve each equation. Then check each answer.

5. $x \div 3 = 6.3$

6. $x \div 0.2 = 3.4$

7. $7x = 4.2$

8. $5x = 4.5$

Original content Copyright © by Holt McDougal. Additions and changes to the original content are the responsibility of the instructor.

Holt McDougal Mathematics

LESSON 3-9

Challenge

Playing Weight

In professional sports, each ball has a maximum, or greatest, weight allowed in play. The maximum official weight for a table tennis ball is only 0.009 ounces. Use the equations below to find the maximum weights, in ounces, of some other sports' balls.

$$\text{(golf)} + 1.601 = \text{(golf)}$$

$$\text{(tennis)} - \text{(golf)} = 0.45$$

$$\text{(8)} - 1.87 = \text{(tennis)} \cdot 2$$

$$\text{(volleyball)} = \text{(8)} + 3.87$$

$$\text{(golf)} + \text{(tennis)} + \text{(volleyball)} + \text{(8)} + 236.03 = \text{(bowling)}$$

Golf Ball	Tennis Ball	Billiard Ball	Volleyball	Bowling Ball
Weight:	Weight:	Weight:	Weight:	Weight:
_______	_______	_______	_______	_______

Original content Copyright © by Holt McDougal. Additions and changes to the original content are the responsibility of the instructor.

     Holt McDougal Mathematics

Problem Solving

LESSON 3-9

Solving Decimal Equations

Write the correct answer.

1. Bee hummingbirds weigh only 0.0056 ounces. They have to eat half their body weight every day to survive. How much food does a bee hummingbird have to eat each day?

2. The desert locust, a type of grasshopper, can jump 10 times the length of its body. The locust is 1.956 inches long. How far can it jump in one leap?

3. In 1900, there were about 1.49 million people living in California. In 2000, the population was 33.872 million. How much did the population grow between 1900 and 2000?

4. Juanita has $567.89 in her checking account. After she deposited her paycheck and paid her rent of $450.00, she had $513.82 left in the account. How much was her paycheck?

Circle the letter of the correct answer.

5. The average body temperature for people is 98.6°F. The average body temperature for most dogs is 3.4°F higher than for people. The average body temperature for cats is 0.5°F lower than for dogs. What is the normal body temperature for dogs and cats?

   A dogs: 101.5°F; cats 102°F

   B dogs: 102°F; cats 101.5°F

   C dogs: 102.5°F; cats 103°F

   D dogs: 102.5°F; cats 102.5°F

6. Seattle, Washington, is famous for its rainy climate. Winter is the rainiest season there. From November through December the city gets an average of 5.85 inches of rain each month. Seattle usually gets 6 inches of rain in December. What is the city's average rainfall in November?

   F 6 inches

   G 5.925 inches

   H 5.8 inches

   J 5.7 inches

7. The equation to convert from Celsius to Kelvin degrees is $K = 273.16 + C$. If it is 303.66°K outside, what is the temperature in Celsius degrees?

   A 576.82°C

   B 30.5°C

   C 305°C

   D 257.68°C

8. The distance around a square mirror is 6.8 feet. Which of the following equations finds the length of each side of the mirror?

   F $6.8 - x = 4$

   G $x \div 4 = 6.8$

   H $4x = 6.8$

   J $6.8 + 4 = x$

Original content Copyright © by Holt McDougal. Additions and changes to the original content are the responsibility of the instructor.

Holt McDougal Mathematics

LESSON 3-9

Reading Strategies
Use a Flowchart

This flowchart can help you work with decimal equations.

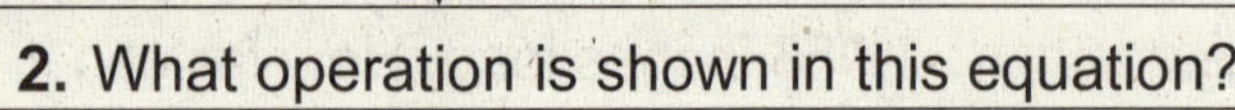

1. How do I read this equation?

↓

2. What operation is shown in this equation?

↓

3. What is the inverse operation for this equation?

Adding or Subtracting Decimals

$x + 9.7 = 15.4$ ⟵ **1.** Read "x plus 9.7 equals 15.4."

$x + \mathbf{9.7} = 15.4$ ⟵ **2.** Addition is shown.

$x + 9.7 - \mathbf{9.7} = 15.4 - \mathbf{9.7}$ ⟵ **3.** The inverse operation is subtraction.

Use this equation to complete Exercises 1–2: $n - 4.5 = 6.3$.

1. Write in words how you read the equation.

____

2. What operation is shown in this equation? What is the inverse of that operation?

____

Use this equation to complete Exercises 3–4: $w + 9 = 4.8$.

3. Write in words how you read this equation.

____

4. What operation is shown in this equation? What is the inverse of that operation?

____

Original content Copyright © by Holt McDougal. Additions and changes to the original content are the responsibility of the instructor.

 Holt McDougal Mathematics

LESSON 3-9 — Puzzles, Twisters & Teasers

Super Cross

The Super Cross is the empty grid below. Fill in the grid with the letters from the equations below to answer the riddle.

First, solve all of the equations. Then enter the letter with the highest value in the top left corner. The next highest goes in the space to its right and then continue filling in the numbers from highest to lowest, left to right and top to bottom.

Read the answer to the riddle by reading down the rightmost column, then down the center column, then down the leftmost column.

$4.9b = 0.637$

$l - 54 = 45.1$

$\dfrac{h}{5} = 6.5$

$i + 9.3 = 15.6$

$4s = 2.8$

$65.7 - t = 24.9$

$a + 4.5 = 21.7$

$\dfrac{e}{5.6} = 3.2$

$34.2s = 3.42$

b	
l	
h	
i	
s	
t	
a	
e	
s	

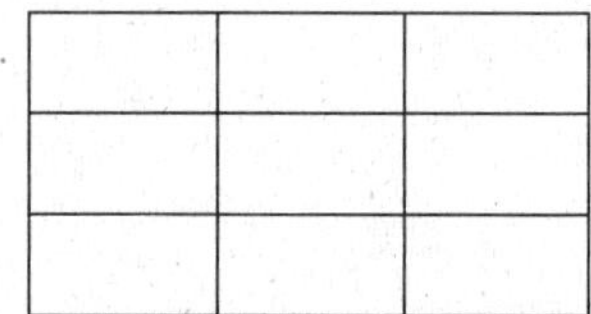

Do you know why the waiter was good at multiplication?

Because he knew

______ ______ ______

______ ______ ______ ______ ______ ______!

Original content Copyright © by Holt McDougal. Additions and changes to the original content are the responsibility of the instructor.

Holt McDougal Mathematics

Answers

LESSON 3-1

Practice A

1. 6 tenths   2. 7 ones
3. 9 hundredths   4. 2 tens
5. 1 tenth   6. 8 hundredths
7. 1 + 0.8; one and eight tenths
8. 3.62; three and sixty-two hundredths
9. 1.52; 1 + 0.5 + 0.02
10. B   11. J
12. July   13. December

Practice B

1. 2 + 0.07; two and seven hundredths
2. 5.007; five and seven thousandths
3. 4.6; 4 + 0.6
4. 16.5; 10 + 6 + 0.5
5. 9.68; nine and sixty-eight hundredths
6. 1 + 0.03 + 0.007; one and thirty-seven thousandths
7. 2.103; two and one hundred three thousandths
8. 0.18; 0.1 + 0.08
9. 6 + 0.1 + 0.01; six and eleven hundredths
10. 3.578; 3.758; 3.875
11. 0.0943; 0.9043; 0.9403
12. 12.75; 12.957; 12.97
13. 1.09; 1.19; 1.9; 1.901
14. the seventh rib
15. the female heart
16. yes   17. Tuesday

Practice C

1. >   2. <
3. <   4. >
5. 12.8692; 12.8962; 12.9682
6. 7.089; 7.098; 8.079; 8.098
7. 6.0521; 6.521; 65.12; 65.21

8. 0.30; 0.304; 0.34; 0.403; 0.43
9. 90.6053; 90.563; 9.653
10. 11.771; 11.717; 11.171; 11.117
11. 8.5; 8.359; 8.3509; 8.0359
12. 2.53; 2.5; 2.35; 2.30; 2.05; 2.03
13. 8.5, 8.54, 8.67, 8.72; in 1988
14. 6.14, 6.03, 6.01, 6.0; 6.14 meters

Review for Mastery

1. The decimal should end with ten thousandths
2. 5 + 0.6 +0.09 + 0.008; five and six hundred ninety-eight thousandths
3. 0 + 0.09 + 0.004; ninety-four thousandths
4. 7.8; seven and eight tenths
5. 0 + 0.1 + 0.02; 0.12
6. <   7. =
8. >   9. >
10. >   11. >
12. 0.43, 0.52, 0.54   13. 3.34, 3.4, 3.43
14. 8.9, 9.5, 9.8   15. 0.083, 0.8, 0.83
16. 0.01, 1.01, 1.1   17. 0.6, 6.0, 6.5

Challenge

Possible answers are given.

1.

Hundreds	Tens	Ones	Tenths
		1	3
		3	1
		5	7
		7	9

Hundredths	Thousandths	Ten-Thousandths
5	7	9
9	5	7
3	9	1
1	3	5

Original content Copyright © by Holt McDougal. Additions and changes to the original content are the responsibility of the instructor.

      Holt McDougal Mathematics

2.

Hundreds	Tens	Ones	Tenths
	8	7	6
	6	8	0
	4	0	8
	2	6	4

Hundredths	Thousandths	Ten-Thousandths
4	0	2
2	7	4
7	2	6
0	8	7

3.

Hundreds	Tens	Ones	Tenths
1	0	2	3
2	1	0	4
3	5	6	2
4	3	5	6

Hundredths	Thousandths	Ten-Thousandths
4	5	6
3	6	5
0	1	4
2	0	1

Problem Solving

1. the blue whale
2. a gray whale
3. a right whale
4. D
5. G
6. D
7. F

Reading Strategies

1. 2 and 17 hundredths
2. 2 + 0.1 + 0.07
3. 3.06
4. 3 + 0.06
5. 1 and 5 tenths
6. 1 + 0.5

Puzzles, Twisters & Teasers

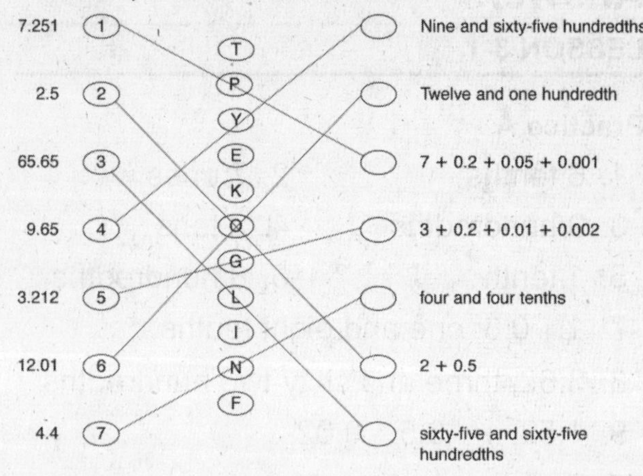

polygon

LESSON 3-2

Practice A

1. 1.8
2. 0.57
3. 13
4. 3.2
5. 24.61
6. 40
7. 4.9
8. 0.85
9. 7.2
10. 6

Possible answers are given.

11. 3
12. 14
13. 6
14. 45
15. 9
16. 15
17. about 2 inches
18. about 4 costumes

Practice B

1. 9.3
2. 13.30
3. 51
4. 1.82

Possible answers are given.

5. 7
6. 72
7. 8
8. 24
9. 9
10. 60

Possible answers are given.

11. from 15 to 17.5
12. from 41 to 43.5
13. about $55.00; about $15.00
14. about 240 miles; about 7 hours

Original content Copyright © by Holt McDougal. Additions and changes to the original content are the responsibility of the instructor.

Holt McDougal Mathematics

Practice C

1. 160
2. 71.83
3. 206.1
4. 105.6

Possible answers are given.

5. 150
6. 10
7. 7
8. 280
9. 12
10. 400

Possible answers are given.

11. from 22 to 24
12. from 69 to 72
13. about 5 inches; 4.6 inches; the estimate is close to the actual data; it is an overestimate.
14. about $600 a week; about $31,200 a year

Review for Mastery

1. 1.0; 9.4; 10.4
2. 2.2; 3.6; 5.8
3. 6.75; 4.26; 2.49
4. 54
5. 77
6. 21
7. 7
8. 64
9. 8
10. 2
11. 70

Possible answers are given.

Challenge

Possible answers are given.

2 People: about $10.00; about $5.00

4 People: about $6.00; about $1.50

4 People: about $16.00; about $4.00

5 People: about $35.00; about $7.00

Problem Solving

Possible answers are given.

1. about 4 years
2. about 82 years
3. about 41.9 kilometers
4. about 5 hours
5. C
6. H
7. B
8. F

Reading Strategies

1. tenths place
2. 35 pounds
3. 42 pounds
4. 42 + 35 = 77

5. $54
6. $22
7. $54 − $22 = $32

Puzzles, Twisters & Teasers

MEET ME AT STAN'S

LESSON 3-3

Practice A

1. 3.8
2. 7.9
2. 3.8
4. 9.2
5. 1.14
6. 16.00
7. 20.0
8. 12.2
9. 8.35
10. 30.7
11. 11.95
12. 2.14
13. B
14. H
15. C
16. F
17. 1.6 meters; 1.8 meters
18. $11.30

Practice B

1. 11.3
2. 3.1
3. 14.19
4. 18.81
5. 4.775
6. 11.56
7. 720.78
8. $722.00
9. 1.9
10. 5.1
11. 2.8
12. 4.18
13. 5.44
14. 1.85
15. $2.71
16. 6.9 yards

Practice C

1. 9.28
2. 34.1
3. 260.765
4. 44.607
5. 0.4641
6. 22.8398
7. 25.527
8. 15.58
9. 12.05
10. 38.802
11. 29.564
12. 49.03
13. 92.92
14. 45.704
15. 3.2046
16. 4
17. 5
18. 3
19. 7
20. 1
21. 4
22. $12.99

Original content Copyright © by Holt McDougal. Additions and changes to the original content are the responsibility of the instructor.

Holt McDougal Mathematics

23. 16.65 feet

Review for Mastery

1. 3.75

2. 0.83

3.

Tens	Ones	Tenths	Hundredths	Thousandths
	4	3		
+		1	4	

5.7

4.

Tens	Ones	Tenths	Hundredths	Thousandths
1	4	4		
−		3	8	

10.6

5.

Tens	Ones	Tenths	Hundredths	Thousandths
	7	3		
+		8	5	

15.8

6.

Tens	Ones	Tenths	Hundredths	Thousandths
1	2	3	4	
−		6	9	

5.44

7. 3.8

8. 33.89

9. 6.95

10. 8.67

Challenge

Possible combinations are given.

1. 2 quarters, 3 dimes,
   1 nickel, 4 pennies

   2 quarters, 2 dimes,
   2 nickels, 9 pennies

   1 quarter, 5 dimes,
   2 nickels, 4 pennies

2. 4 quarters, 2 dimes,
   1 nickel, 3 pennies

   2 quarters, 4 dimes,
   5 nickels, 13 pennies

   3 quarters, 4 dimes,
   2 nickels, 3 pennies

3. 1 quarter, 3 dimes,
   1 nickel, 5 pennies

   1 quarter, 1 dime,
   4 nickels, 10 pennies

   1 quarter, 2 dimes,
   3 nickels, 5 pennies

4. 8 quarters, 2 dimes,
   1 nickel, 5 pennies

   4 quarters, 10 dimes,
   4 nickels, 10 pennies

   6 quarters, 3 dimes,
   4 nickels, 30 pennies

Problem Solving

1. 4.65 million tons

2. 108.55 million tons

3. B

4. H

5. A

6. G

Original content Copyright © by Holt McDougal. Additions and changes to the original content are the responsibility of the instructor.

Holt McDougal Mathematics

Reading Strategies

1. It helps you line up decimal points.

2.

Ones	Tenths	Hundredths
3.	2	5
1.	0	6
+ 2.	9	0

3.

Tens	Ones	Tenths	Hundredths
2	3.	8	2
−	7.	2	0

4. 7.21

5. 16.62

6. 2.9 written as 2.90; 7.2 written as 7.20

Puzzles, Twisters & Teasers

First Word Cost: $10.87; $4.13 remaining; Missing Letter: D

Second Word Cost: $10.41; $4.59 remaining; Missing Letter: L

FOLD it in HALF

LESSON 3-4

Practice A

1. 26,700

2. 3,810

3. 192

4. B

5. H

6. 3

7. 4

8. 2

9. 4.7

10. 6.58

11. 8.9

12. 340

13. 79,000

14. 1,750

15. 124,000

16. 960,000

17. 1,280,000

18. $1.1 \cdot 10^4$ pounds

Practice B

1. 34,500

2. 6,520

3. 1,840

4. $1.67 \cdot 10^4$

5. $4.68 \cdot 10^3$

6. $5.834 \cdot 10^7$

7. 32,500

8. 7,080,000

9. 12,090,000

10. 680,000,000

11. 51,000

12. 6

13. C

14. A

15. C

16. B

17. 12,300,000; $1.23 \cdot 10^7$

18. 12,060,000

Practice C

1. 1,670

2. 9,360

3. 35,500

4. $6.389 \cdot 10^6$

5. $1.052 \cdot 10^8$

6. $1.52 \cdot 10^8$

7. 15,089

8. 251,600,000

9. 17,711,000

10. 3,960,400

11. 2,840

12. 8.69

13. $2.5 \cdot 10^5$

14. $6.5 \cdot 10^1$

15. $8.9 \cdot 10^4$

16. $1.54 \cdot 10^8$

17. $7.3 \cdot 10^2$

18. $1.024 \cdot 10^7$

19. the United States; 104,491,000 more people

20. 2,270,000,000 passengers

Review for Mastery

1. $1.75 \cdot 10^5$

2. $2.98 \cdot 10^2$

3. $5.764 \cdot 10^3$

4. $8.3 \cdot 10^1$

5. $4.03 \cdot 10^4$

6. $2 \cdot 10^6$

7. $5.101 \cdot 10^4$

8. $1.90025 \cdot 10^5$

9. 5,620

10. 723.8

11. 990,000

12. 65.3

13. 53,600

14. 240

15. 4,350

16. 800,000

17. 10,000

18. 2030

19. 112

20. 3,002,000

Original content Copyright © by Holt McDougal. Additions and changes to the original content are the responsibility of the instructor.

Holt McDougal Mathematics

Challenge

1. 92,900,000	2. 483,600,000
3. 141,600,000	4. 36,000,000
5. 2,794,000,000	6. 3,675,000,000
7. 887,000,000	8. 1,784,000,000
9. 67,200,000	

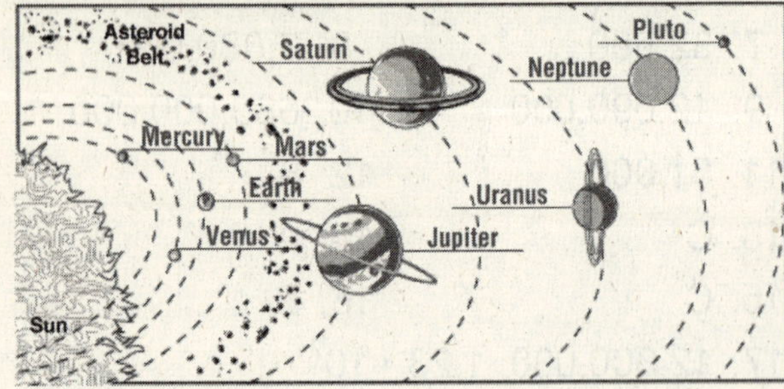

Problem Solving

1. $8.742 \cdot 10^5$
2. 14,000,000,000 years
3. $2.47 \cdot 10^8$ people
4. 468,800,000 tons

5. B	6. H
7. D	8. H

Reading Strategies

1. standard form
2. scientific notation
3. standard form
4. words and symbols
5. scientific notation

Puzzles, Twisters & Teasers

company folded

LESSON 3-5

Practice A

1. 0.08	2. 0.12
3. 0.6	4. 0.99
5. 1.25	6. 4.2
7. 0.2	8. 1.8
9. 0.51	10. 2.68
11. 1.92	12. 0.64
13. 0.2	14. 1.0

15. 1.8	16. 2.4
17. 3.4	18. 4.8
19. 7.5 pounds	20. $11.20

Practice B

1. 0.21	2. 0.02
3. 0.16	4. 0.7
5. 0.121	6. 8.1
7. 0.0232	8. 3.225
9. 1.384	10. 12.034
11. 1.584	12. 0.0063
13. 4	14. 18.4
15. 5.92	16. 24.96
17. 4.696	18. 112.64

19. 28.8 kilometers; 129.6 kilometers

20. $8.28

Practice C

1. 2.958	2. 5.5246
3. 0.2115	4. 2.6795
5. 0.90798	6. 33.18855
7. 0.007482	8. 153.20232
9. 0.40344	10. 32.3
11. 0.085	12. 107.219
13. 190.655	14. 35.003
15. 1,289.348	16. 5.346
17. 151.03	18. 341.92
19. 6.4	20. 247.7 days
21. 687 days	

Review for Mastery

1. 0.69	2. 0.82
3. 0.055	4. 0.64
5. 0.45	6. 0.84
7. 0.32	8. 0.88
9. 0.16	10. 0.63
11. 0.25	12. 0.18
13. 0.1	14. 0.16
15. 0.09	16. 0.28

Original content Copyright © by Holt McDougal. Additions and changes to the original content are the responsibility of the instructor.

Holt McDougal Mathematics

Challenge

1 Week: 17.5 cm; 0.063 mm; 1.96 in;

0.25 Day: 0.35 mm; 0.000625 mm;
    0.0295 in.

Problem Solving

1. $30.00

2. $40.00

3. $57.00

4. about 3 times

5. B

6. J

7. C

8. G

Reading Strategies

1.

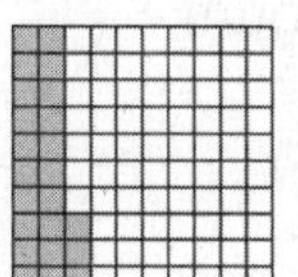 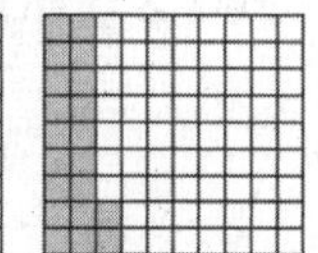 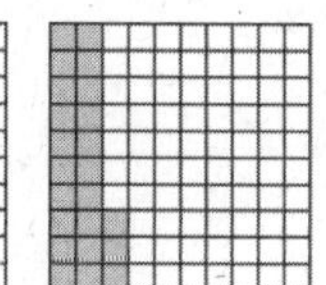 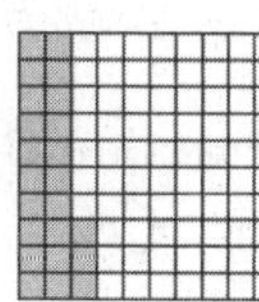

2. 0.23 + 0.23 + 0.23 + 0.23

3. 0.92

4. 4 × 0.23

5. 0.92

Puzzles, Twisters & Teasers

0.0512; 0.072; 0.021375; 0.019;
Makiko

LESSON 3-6

Practice A

1. 0.7

2. 0.9

3. 0.6

4. 0.8

5. 0.05

6. 0.6

7. 0.2

8. 0.3

9. 0.08

10. 0.6

11. 0.7

12. 0.1

13. 0.3

14. 1.2

15. 0.8

16. 0.6

17. 0.4

18. 0.2

19. $0.70

20. 2.4 inches

Practice B

1. 0.09

2. 0.46

3. 1.2

4. 1.7

5. 0.91

6. 3.7

7. 1.73

8. 0.012

9. 32.5

10. 0.0005

11. 0.355

12. 6.3

13. 1.047

14. 0.5235

15. 0.1745

16. 0.1047

17. 0.1396

18. 0.0698

19. 526.38 pounds

20. $22.62

Practice C

1. 0.295

2. 0.0496

3. 12.109

4. 0.947

5. 9.65

6. 0.056

7. 1.605

8. 0.269

9. 0.003744

10. 0.8

11. 0.09

12. 0.004

13. 0.18

14. 0.07

15. 0.0002

16. 1.5

17. 7.47

18. 63.91

19. 16.94

20. 5.6284 million

21. Kim

Review for Mastery

1. 0.06

2. 0.04

3. 0.05

4. 0.14

5. 0.12

6. 0.13

7. 0.09

8. 0.08

Challenge

$0.69; $0.45; $0.40
    12 for $4.80

$0.75; $0.85; $0.81
    1 pound for $0.75

$0.33; $0.31; $0.33
    12-ounce box for $3.72

$0.18; $0.22; $0.17
    24-pack for $4.08

Problem Solving

1. $8.35

2. 0.6 milligrams

3. 2.9 feet

4. 1.85 meters

5. B

6. H

Original content Copyright © by Holt McDougal. Additions and changes to the original content are the responsibility of the instructor.

Holt McDougal Mathematics

7. A   8. H

Reading Strategies

1.
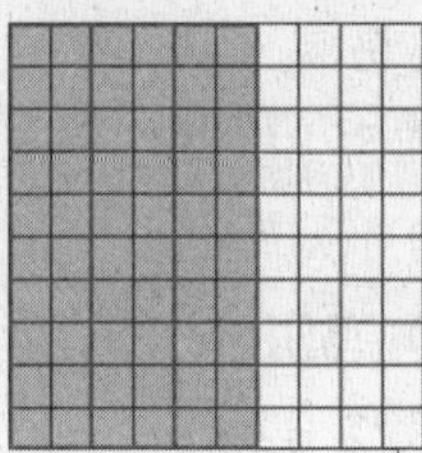

2.
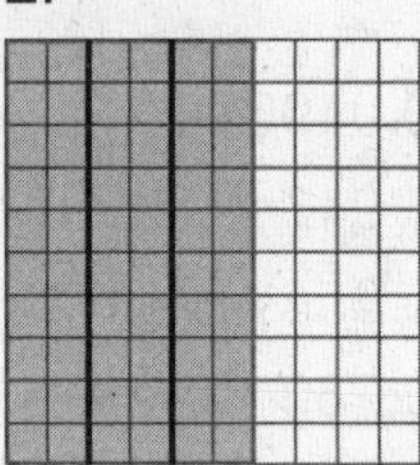

3. 0.20

4. 0.60 ÷ 3 = 0.20

Puzzles, Twisters & Teasers

MONEY PER STATE
$38.69
$213.08
$89.06
$47.28
$17.88
$38.64
$294.93

Ohio

LESSON 3-7

Practice A

1. 6   2. 7
3. 8   4. 9
5. 6   6. 8
7. 11   8. 2
9. 6   10. 5
11. 12   12. 9
13. 24   14. 12
15. 16   16. 8
17. 6   18. 4

19. 4 pounds   20. 3 months

Practice B

1. 10   2. 8
3. 4.7   4. 7
5. 3.2   6. 7.5
7. 4   8. 15
9. 8.4   10. 2.4
11. 3.4   12. 20.5
13. 34   14. 1.5
15. 9   16. 12
17. 3.6   18. 0.05
19. 9 weeks   20. 2.5 hours

Practice C

1. 1.9   2. 7.4
3. 8   4. 32.6
5. 25   6. 8.4
7. 2.089   8. 19.06
9. 6.047   10. 78.25
11. 39.125   12. 31.3
13. 19.5625   14. 9.78125
15. 6.26   16. 3.46
17. 4.6   18. 7.87
19. 46.53   20. 2.4 and 1.5
21. 3.5 and 1.8

Review for Mastery

1. 0.6   2. 5
3. 0.25   4. 15
5. 0.2   6. 0.7
7. 50   8. 0.8
9. 0.2   10. 5
11. 700   12. 0.02

Challenge

1. Students should draw 6 lines across the cloth to cut it into 7 equal pieces.

2. Students should draw 4 lines across the cloth to cut it into 5 equal pieces.

Original content Copyright © by Holt McDougal. Additions and changes to the original content are the responsibility of the instructor.

Holt McDougal Mathematics

3. Students should draw 5 lines across the cloth to cut it into 6 equal pieces.

4. Students should draw 7 lines across the cloth to cut it into 8 equal pieces.

5. Students should draw 3 lines across the cloth to cut it into 4 equal pieces.

Problem Solving

1. 15 feet
2. 55.8 miles per hour
3. 27.6 miles per gallon
4. 169 hamburgers
5. C
6. H
7. C
8. F

Reading Strategies

1. 0.002
2. 200,000
3. 400 ÷ 0.002 = 200,000
4. 0.03
5. 30,000
6. 900 ÷ 0.03 = 30,000

Puzzles, Twisters & Teasers

Price per Oz
$0.56
$0.30
$0.15
$0.50
$0.10
$0.61
$0.12

A NERVOUS WRECK

LESSON 3-8

Practice A

1. A
2. G
3. A
4. J
5. $3.10
6. 4 vans
7. 7 packs
8. 8 packs

Practice B

1. B
2. H
3. A
4. H
5. 6 costumes
6. 11 weeks
7. 6 costumes
8. $2.19

Practice C

1. 9.5 gallons
2. 5 rolls of the 24-photo or 6 rolls of the 18-photo film
3. 2 times
4. $1.20 per gallon
5. $1.25
6. 60.3 miles per hour
7. $4.21
8. 54 minutes

Review for Mastery

1. 1.5 cups are needed for each serving
2. Marla needs 4 cases.
3. Kenny can play 8 games.

Challenge

Number of Packs to Buy	Number of Left Over Items	Total Price of Items
4	2	$7.40
2	20	$5.60
6	0	$11.70
1	15	$3.65
3	6	$6.15
Grand Total Price:		$34.50
Cost Per Student:		$1.15

Original content Copyright © by Holt McDougal. Additions and changes to the original content are the responsibility of the instructor.

Holt McDougal Mathematics

Problem Solving

1. $3.35
2. 6 vans
3. 5 necklaces
4. $0.57
5. B
6. J
7. A
8. H

Reading Strategies

1. round down; You can make only 6 bows with 50 inches of ribbon.
2. round up; You need another lunch table to seat everyone.

Puzzles, Twisters & Teasers

10.66; 9.5; 11.66; 12.16

# of People John should invite: 9

LETTER BANK

U   R

R

O   M

RUMOR

LESSON 3-9

Practice A

1. B
2. H
3. $a = 1.7$; $1.7 - 0.4 = 1.3$
4. $n = 0.9$; $2 \cdot 0.9 = 1.8$
5. $x = 0.5$; $0.8 + 0.5 = 1.3$
6. $p = 2.8$; $2.8 \div 4 = 0.7$
7. $w = 3.5$; $3.5 - 2.3 = 1.2$
8. $q = 0.5$; $0.5 \cdot 0.5 = 0.25$
9. $s = 0.3$; $5.7 + 0.3 = 6.0$
10. $b = 1.8$; $1.8 \div 3 = 0.6$
11. $t = 4.7$; $4.7 - 3.1 = 1.6$
12. $y = 0.5$; $3 \cdot 0.5 = 1.5$
13. $a = 6.3$ square inches
14. $4.50

Practice B

1. $a = 7.5$; $7.5 - 2.7 = 4.8$
2. $b = 13.3$; $13.3 \div 7 = 1.9$
3. $w = 10.3$; $10.3 - 6.5 = 3.8$
4. $p = 0.68$; $0.68 \div 0.4 = 1.7$
5. $x = 3.5$; $4.5 + 3.5 = 8$
6. $b = 7.5$; $7.5 \div 3 = 2.5$
7. $s = 7.4$; $7.8 + 7.4 = 15.2$
8. $q = 6$; $1.63 \cdot 6 = 9.78$
9. $x = 2.01$; $0.05 + 2.01 = 2.06$
10. $n = 1.4$; $1.7 \cdot 1.4 = 2.38$
11. $t = 18.67$; $18.67 - 6.08 = 12.59$
12. $q = 1.8$; $9 \cdot 1.8 = 16.2$
13. $w = 19.2$; $19.2 - 8.9 = 10.3$
14. $n = 2.3$; $1.4 \cdot 2.3 = 3.22$
15. $t = 13.5$; $13.5 - 12.7 = 0.8$
16. $a = 2.7$; $3.8 + 2.7 = 6.5$
17. 3.2 centimeters
18. $2.50

Practice C

1. $a = 12.589$; $12.589 - 0.089 = 12.5$
2. $b = 13.16$; $13.16 \div 2.8 = 4.7$
3. $w = 28.56$; $28.56 - 12.76 = 15.8$
4. $p = 0.3332$; $0.3332 \div 0.07 = 4.76$
5. $x = 0.2284$; $1.7806 + 0.2284 = 2.009$
6. $b = 63.05$; $63.05 \div 6.5 = 9.7$
7. $s = 7.7056$; $8.3044 + 7.7056 = 16.01$
8. $q = 85$; $0.009 \cdot 85 = 0.765$
9. $x = 6.736$; $23.764 + 6.736 = 30.5$
10. $n = 2.3$; $1.7 \cdot 2.3 + 3.8 = 7.71$
11. $t = 64.86$; $64.86 - 64.36 = 0.5$
12. $q = 3$; $9.2 \cdot 3 = 27.6$
13. $w = 1.9$; $(8.4 \cdot 1.9) - 0.67 = 15.29$
14. $n = 15$; $(1.4 \cdot 15) + 64 = 85$
15. $13.37; 2.8 pounds
16. 28.47 square feet; 14.235 square feet

Original content Copyright © by Holt McDougal. Additions and changes to the original content are the responsibility of the instructor.

Holt McDougal Mathematics

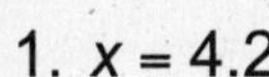

Review for Mastery

1. $x = 4.2$ 2. $x = 8.2$

3. $x = 10.45$ 4. $x = 3.1$

5. $x = 18.9$ 6. $x = 0.68$

7. $x = 0.6$ 8. $x = 0.9$

Challenge

1.61 ounces; 2.06 ounces; 5.99 ounces;
9.86 ounces; 255.55 ounces

Problem Solving

1. 0.0028 ounces   2. 19.56 inches

3. by 32.382 million   4. $395.93

5. B   6. J

7. B   8. H

Reading Strategies

1. n minus 4.5 equals 6.3.

2. subtraction; addition

3. w plus 9 equals 4.8.

4. addition; subtraction

Puzzles, Twisters & Teasers

l	t	h
e	a	i
s	b	s

Because he knew **his tables**

Original content Copyright © by Holt McDougal. Additions and changes to the original content are the responsibility of the instructor.

Holt McDougal Mathematics